SERVSAFE® ESSENTIALS

FIFTH EDITION

Updated with the *2009 FDA Food Code*

NATIONAL
RESTAURANT
ASSOCIATION®

Disclaimer

Introduction

Unit I   The Food Safety Challenge

Unit II   The Flow of Food Through the Operation

Unit III   Food Safety Management Systems, Facilities, and Pest Management

Unit IV   Food Safety Regulations and Employee Training

Appendix

Index

A Message from
The National Restaurant Association

The National Restaurant Association is pleased to bring you the fifth edition of *ServSafe® Essentials*, updated with the *2009 FDA Food Code*.

By opening this book, you are taking the first step in your commitment to food safety. ServSafe training introduces you to the basic information you need to know to serve safe food. Training also helps you understand all the food safety risks faced by your operation. Once you're aware of these risks, you can find ways to reduce them, which will help you keep your operation, your customers, and your employees safe.

The ServSafe Training and Certification Program provides you with the knowledge, skills, and abilities you need to do your job. It leads the way in setting high food safety standards.

Created by Foodservice Industry Leaders　　You can be confident knowing the ServSafe program was created by the foodservice industry, for the foodservice industry. Those who deal with the same food safety issues you face every day determined the topics you will learn in this book. From the basics of handwashing to more complex topics such as foodborne pathogens, your industry peers have provided you with the building blocks to keep food safe throughout your operation.

Delivered by Certified ServSafe Instructors　　Your success in learning is important to us. That's why the National Restaurant Association has implemented higher standards for the people who train you, our Certified ServSafe Instructors.

Performed and Reinforced by You　　Food safety doesn't stop once you've completed your ServSafe training and certification. It's only just begun. Training is a process, not an event. It is now your responsibility to take the knowledge you learned and share it with your employees. The information in chapter 14, Employee Food Safety Training, will help you find ways to train your staff. ServSafe also offers training materials to help you teach key food safety topics to hourly employees. Free materials are available in the Resource Center on *www.ServSafe.com*.

Thank you for investing your time in ServSafe training. We view your training as a critical piece of your success, and we are confident that you'll benefit greatly by applying what you learn to your own operation. For more information on all ServSafe programs, visit *www.ServSafe.com*.

About the National Restaurant Association

The National Restaurant Association, founded in 1919, is the leading business association for the restaurant industry, which is comprised of 935,000 restaurant and foodservice outlets and a workforce of 12.8 million employees—making it the cornerstone of the economy, career opportunities, and community involvement. Along with the National Restaurant Association Educational Foundation, NRA Solutions and the Association work to represent, educate, and promote the rapidly growing industry. For more information, visit our Web site at *www.restaurant.org*.

International Food Safety Council®

The International Food Safety Council's mission is to heighten the awareness of the importance of food safety education throughout the restaurant and foodservice industry. The council envisions a future in which foodborne illness no longer exists.

For more information about the International Food Safety Council, sponsorship opportunities, and initiatives, please visit *www.ServSafe.com.*

Active Founding Sponsors

American Egg Board

The Beef Checkoff

Ecolab Inc.

SYSCO Corporation

Campaign Sponsors

Cintas Corporation

Rubbermaid Commercial Products

Acknowledgements

The development of the *ServSafe Essentials* text would not have been possible without the expertise of our many advisors and manuscript reviewers. Thanks to the following organizations for their time, effort, and dedication to creating this fifth edition.

3M

Boskovich Farms, Inc.

Centers for Disease Control and Prevention

Comark Instruments

The Cooking and Hospitality Institute of Chicago

Cooper-Atkins Corporation

Daydots

Fluke Corporation

Kendall College School of Culinary Arts, Chicago

Orkin Commercial Services

Washburne Culinary Institute, Chicago

How to Use *ServSafe Essentials*

The plan below will help you study and remember the food safety principles in this book.

Units

ServSafe Essentials is broken into four units. Each unit has an introduction page that shows you the chapters in that unit. It also shows you icons for the major topics that are in each chapter. Below is an example of the icons for unit 1.

Foodborne Illnesses | Preventing Foodborne Illnesses | Pathogens | Specific Foodborne Pathogens | Biological Toxins | Contamination | Food Allergens | How Foodhandlers Can Contaminate Food | A Good Personal Hygiene Program

You will also see these icons throughout each chapter of the unit. They appear in three locations in each chapter.

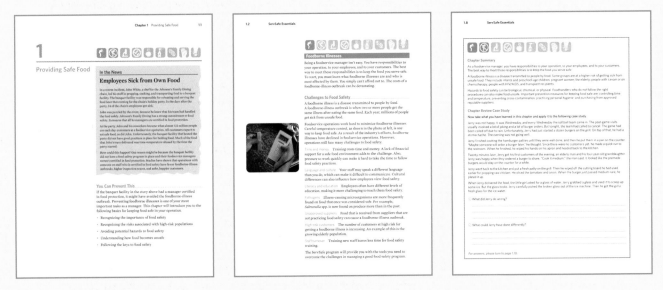

Opening page   All the icons for the unit are at the top of the page. The icons for the major topics that are covered in the chapter are highlighted.

Beginning of a major topic   All the icons for the unit are at the top of the page. The highlighted icon is for the major topic covered in that section.

Chapter summary and review activities page   All the icons for the unit are at the top of the page. The icons for the major topics covered in the chapter are highlighted.

Beginning Each Chapter

Before you start reading each chapter, you can prepare by reviewing these sections.

In the News   This real-world story introduces you to the chapter. It shows you how practicing food safety the right way or the wrong way can positively or negatively impact an operation. What happens in the story relates to the concepts you will learn in the chapter.

You Can Build On This   This section tells you the positive things that happened in the "In the News" story and tells you what you will learn in the chapter. These topics are the essential practices for keeping food in your operation safe.

You Can Prevent This   This section tells you the negative things that happened in the "In the News" story and tells you how the situation could have been prevented. Additionally, it lists what you will learn in the chapter. These topics are the essential practices for keeping food in your operation safe.

Concepts from Earlier Chapters   These concepts are important definitions or explanations that you learned earlier in the book. They are important for understanding the content in the chapter.

Throughout Each Chapter

Use the following learning tools to help you identify and learn key food safety principles as you read each chapter.

Photos   Photos give you visual examples of key principles in the book. They are either in the margin next to a principle or in the text itself. Some of the photos show you what you should do, while others show you what you shouldn't do. If the photo is a positive practice, it has a ✔ in the left corner. If the photo is a negative practice, it has a ✘ in the left corner, meaning that this practice should never be done.

Charts, illustrations, and tables   These visuals either present or organize content so that it is easier for you to learn. In some instances, they are used to reinforce key principles in the book.

Pathogen prevention icons   These icons point to actions that you can take to prevent a specific pathogen from making people sick.

> **Apply Your Knowledge activities**   These activities allow you to apply the key food safety practices in a chapter. At the end of each major topic, you can practice what you learned. Answers are at the end of each chapter.

> **Something to Think About...**   Many food safety stories appear in this book. Some of these real-world stories focus on foodborne illnesses that happened because food was not handled safely. They show the importance of following food safety practices. They also allow you to apply what you have learned by asking how the illness could have been prevented. Other stories show you real-world solutions to food safety problems that operations have experienced. These solutions can help you improve the food safety practices in your own operation.

> **How This Relates to Me**   Some of the food safety practices in this book may differ from your local laws. To help you remember these differences, you can record your local regulatory requirements in these write-in areas of the book.

At the End of Each Chapter

The end of each chapter gives you three opportunities to review the content you just learned.

Chapter summary   A summary is at the end of each chapter to help remind you of the major topics you learned.

Chapter review case studies   These food safety case studies ask you to identify the errors made by the foodhandlers in each story and the practices that the staff should have followed.

Study questions   These multiple-choice questions are based on key food safety principles. If you have trouble answering them, you should review the content again. Answers are at the end of each chapter.

Implementing the Food Safety Practices You Learned in the ServSafe Program

The ServSafe program will give you the information you need to keep food safe in your operation. It is your responsibility to put that information into practice. To do this, you must take what you have learned and use it to examine the following parts of your operation.

- Current food safety policies and procedures

- Employee training

- Your facility

The steps listed below will help you make the comparison that will take you from where you are today to where you need to be to *consistently* keep food safe in your operation.

1 **Evaluate your current food safety practices using the Food Safety Evaluation Checklist in the appendix.** This checklist, which begins on page A.2, identifies the most critical food safety practices an operation must follow. It is a series of Yes/No questions that will help you see areas for improvement. When you have checked a "No" in this checklist, you have found a gap in your food safety practices. These gaps are the starting point for improving your current food safety program.

2 **Review the "How This Relates to Me" areas throughout *ServSafe Essentials*.** These are the write-in areas in the book that help you remember the food safety practices required by your local regulatory authority. If a requirement is different from your company policy or is not addressed by it, you have found a gap in your food safety program and a chance to make an improvement.

3 **Determine the cause of the gaps you found in steps 1 and 2.** For example, if you find that your walk-in cooler cannot hold food at 41°F (5°C) or lower, you have found a gap. There are many things that could have caused this gap, including faulty equipment, a walk-in door that is opened too often, etc. You must look at each of these potential causes to determine the true reason for the gap.

4 **Create a solution that closes the gaps.** Your solution might include any of the following tasks.

- Developing or revising standard operating procedures (SOPs)

- Making improvements to current equipment or buying new equipment

- Training or retraining staff

5 **Evaluate your solution regularly to make sure it has closed the gaps you found in steps 1 and 2.**

The Food Safety Challenge

1

Providing Safe Food

In the News

Employees Sick from Own Food

In a recent incident, John White, a chef for the Johnson's Family Dining chain, led his staff in prepping, cooking, and transporting food to a banquet facility. The banquet facility was responsible for reheating and serving the food later that evening for the chain's holiday party. In the days after the party, 34 of the chain's employees got sick.

John was puzzled by the event, because he knew that his team had handled the food safely. Johnson's Family Dining has a strong commitment to food safety. It ensures that all its managers are certified in food protection.

At the party, John and his coworkers became what almost 132 million people are each day: customers at a foodservice operation. All customers expect to eat safe food, as did John. Unfortunately, the banquet facility that hosted the party did not have good practices in place for holding food. Much of the food that John's team delivered was time-temperature abused by the time the party started.

How could this happen? One reason might be because the banquet facility did not have a food safety program in place and their foodservice managers weren't certified in food protection. Studies have shown that operations with someone on staff who is certified in food safety have fewer foodborne-illness outbreaks, higher inspection scores, and safer, happier customers.

You Can Prevent This

If the banquet facility in the story above had a manager certified in food protection, it might have avoided the foodborne-illness outbreak. Preventing foodborne illnesses is one of your most important tasks as a manager. This chapter will introduce you to the following basics for keeping food safe in your operation.

- Recognizing the importance of food safety

- Recognizing the risks associated with high-risk populations

- Avoiding potential hazards to food safety

- Understanding how food becomes unsafe

- Understanding important prevention measures for keeping food safe

Foodborne Illnesses

Being a foodservice manager isn't easy. You have responsibilities to your operation, to your employees, and to your customers. The best way to meet those responsibilities is to keep the food you serve safe. To start, you must learn what foodborne illnesses are and who is most affected by them. You simply can't afford not to. The costs of a foodborne-illness outbreak can be devastating.

Challenges to Food Safety

A foodborne illness is a disease transmitted to people by food. A foodborne-illness outbreak is when two or more people get the same illness after eating the same food. Each year, millions of people get sick from unsafe food.

Foodservice operations work hard to minimize foodborne illnesses. Careful temperature control, as shown in the photo at left, is one way to keep food safe. As a result of the industry's efforts, foodborne illnesses have declined in foodservice operations. However, operations still face many challenges to food safety.

Time and money   Training costs time and money. A lack of financial support for a safe food environment adds to the challenge. Also, pressure to work quickly can make it hard to take the time to follow food safety practices.

Language and culture   Your staff may speak a different language than you do, which can make it difficult to communicate. Cultural differences can also influence how employees view food safety.

Literacy and education   Employees often have different levels of education, making it more challenging to teach them food safety.

Pathogens   Illness-causing microorganisms are more frequently found on food that once was considered safe. For example, *Salmonella* spp. is now found on produce more than in the past.

Unapproved suppliers   Food that is received from suppliers that are not practicing food safety can cause a foodborne-illness outbreak.

High-risk customers   The number of customers at high risk for getting a foodborne illness is increasing. An example of this is the growing elderly population.

Staff turnover   Training new staff leaves less time for food safety training.

The ServSafe program will provide you with the tools you need to overcome the challenges in managing a good food safety program.

The Cost of Foodborne Illnesses

Foodborne illnesses cost the United States billions of dollars each year. National Restaurant Association figures show that one foodborne-illness outbreak can cost an operation thousands of dollars and even result in closure.

Costs of a Foodborne Illness to an Operation

Loss of customers and sales

Loss of reputation

Negative media exposure

Lowered staff morale

Lawsuits and legal fees

Staff missing work

Increased insurance premiums

Staff retraining

Most important are the human costs. Victims of foodborne illnesses may experience lost work, medical costs, long-term disability, and even death.

Populations at High Risk for Foodborne Illnesses

Certain groups of people have a higher risk of getting a foodborne illness.

Elderly people

People's immune systems weaken with age. The immune system is the body's defense against illness.

Infants and preschool-age children

Very young children have not built up strong immune systems.

Pregnant women

Women's immune systems are compromised during pregnancy.

Other populations

- People with cancer or on chemotherapy
- People with HIV/AIDS
- Transplant recipients

Apply Your Knowledge

Who's at Risk?

Place a ✔ next to the people in high-risk populations.

① _____ 68-year-old man

② _____ 25-year-old man on chemotherapy

③ _____ 23-year-old pregnant woman

④ _____ 41-year-old man on blood-pressure medication

⑤ _____ 45-year-old man

⑥ _____ 16-year-old girl

⑦ _____ 38-year-old transplant recipient

⑧ _____ 3-year-old girl

For answers, please turn to page 1.10.

Preventing Foodborne Illnesses

To prevent foodborne illnesses, you must recognize the hazards that can make food unsafe. These hazards can come from pathogens, chemicals, or objects. They might also come from certain unsafe practices in your operation. Most of these hazards can be controlled by focusing on personal hygiene, time and temperature control, and cross-contamination.

Potential Hazards to Food Safety

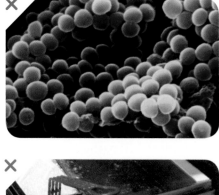

Unsafe food is usually the result of contamination, which is the presence of harmful substances in the food. Some food safety hazards are caused by humans or by the environment. Others can occur naturally.

Potential hazards to food safety are divided into three categories.

Biological   Pathogens are the greatest threat to food safety. They include certain viruses, parasites, fungi, and bacteria, like the photo of *Staphylococcus aureus* shown at left. Some plants, mushrooms, and seafood that carry harmful toxins (poisons) are also included in this group.

Chemical   Foodservice chemicals can contaminate food if they are used incorrectly. The photo at left shows one example of how chemicals may contaminate food. This group also includes cleaners, sanitizers, polishes, machine lubricants, and toxic metals that leach from cookware into food.

Physical   Foreign objects like hair, dirt, bandages, metal staples, or broken glass can get into food. The photo at left shows this type of physical hazard. Naturally occurring objects, like fish bones in fillets, are also physical hazards.

Each of the hazards listed above is a danger to food safety. But the greatest threat to an operation's food safety program is biological hazards. Pathogens are responsible for most foodborne-illness outbreaks.

How Food Becomes Unsafe

The Centers for Disease Control and Prevention (CDC) has identified the five most common risk factors that cause foodborne illness.

❶ Purchasing food from unsafe sources

❷ Failing to cook food adequately

❸ Holding food at incorrect temperatures

❹ Using contaminated equipment

❺ Practicing poor personal hygiene

Except for purchasing food from unsafe sources, each cause listed above is related to three main factors. These are time-temperature abuse, cross-contamination, and poor personal hygiene.

Time-temperature abuse   Food has been time-temperature abused when it has stayed too long at temperatures that are good for the growth of pathogens. A foodborne illness can result if food is time-temperature abused, which can happen in many ways.

- Food is not held or stored at the right temperature, as shown in the photo at left.
- Food is not cooked or reheated enough to kill pathogens.
- Food is not cooled the right way.

Cross-contamination   Pathogens can be transferred from one surface or food to another. Cross-contamination can cause a foodborne illness in many ways.

- Contaminated ingredients are added to food that receives no further cooking.
- Ready-to-eat food touches contaminated surfaces.
- Contaminated food touches or drips fluids onto cooked or ready-to-eat food, as shown in the photo at left.
- A foodhandler touches contaminated food and then touches ready-to-eat food.
- Contaminated cleaning towels touch food-contact surfaces.

Poor personal hygiene   Foodhandlers can cause a foodborne illness if they do any of the following actions.

- Fail to wash their hands the right way after using the restroom or after any time their hands get dirty
- Come to work while sick
- Cough or sneeze on food, as shown in the photo at left
- Touch or scratch wounds, and then touch food

Important Prevention Measures

Now that you know how food can become unsafe, you can use this knowledge to keep food safe. Focus on these measures.

- Controlling time and temperature
- Preventing cross-contamination
- Practicing personal hygiene
- Purchasing from approved, reputable suppliers

Set up standard operating procedures that focus on these areas. The ServSafe program will show you how to design these procedures in later chapters.

As a manager, your job is more than just understanding food safety practices. You also have to train the employees in your operation, as shown in the photo at left. Most important, you must then monitor them to make sure they follow the procedures.

Apply Your Knowledge

What's the Problem?

Six dangerous actions are listed below. Under each example, write an ✖ next to the option that best describes how the food became unsafe.

① A package of raw chicken breasts is left out at room temperature.

_____ Time-temperature abuse   _____ Poor personal hygiene   _____ Cross-contamination

② A foodhandler sneezes on a salad.

_____ Time-temperature abuse   _____ Poor personal hygiene   _____ Cross-contamination

③ A foodhandler cooks a rare hamburger.

_____ Time-temperature abuse   _____ Poor personal hygiene   _____ Cross-contamination

④ A foodhandler scratches a cut, and then continues to make a sandwich.

_____ Time-temperature abuse   _____ Poor personal hygiene   _____ Cross-contamination

⑤ A foodhandler leaves the restroom without washing her hands.

_____ Time-temperature abuse   _____ Poor personal hygiene   _____ Cross-contamination

⑥ A foodhandler cuts up raw chicken. He then uses the same knife to chop carrots for a salad.

_____ Time-temperature abuse   _____ Poor personal hygiene   _____ Cross-contamination

For answers, please turn to page 1.10.

Chapter Summary

As a foodservice manager, you have responsibilities to your operation, to your employees, and to your customers. The best way to meet those responsibilities is to keep the food you serve safe.

A foodborne illness is a disease transmitted to people by food. Some groups are at a higher risk of getting sick from unsafe food. They include infants and preschool-age children; pregnant women; the elderly; people with cancer or on chemotherapy; people with HIV/AIDS, and transplant recipients.

Hazards to food safety can be biological, chemical, or physical. Foodhandlers who do not follow the right procedures can also make food unsafe. Important prevention measures for keeping food safe are: controlling time and temperature; preventing cross-contamination; practicing personal hygiene; and purchasing from approved, reputable suppliers.

Chapter Review Case Study

Now take what you have learned in this chapter and apply it to the following case study.

Jerry was not happy. It was Wednesday, and every Wednesday the softball team came in. The post-game visits usually involved a lot of joking and a lot of burger orders. But tonight, the team had called to cancel. The game had been called off due to rain. Unfortunately, Jerry had just started a dozen burgers on the grill. On top of that, he had a stomachache. The evening was not going well.

Jerry finished cooking the hamburger patties until they were well done, and then he put them in a pan on the counter. "Maybe someone will order a burger later," he thought. Since there were no customers yet, he made a quick run to the restroom. When he finished, he wiped his hands on his apron and headed back to the kitchen.

Twenty minutes later, Jerry got his first customers of the evening, an elderly man and his four-year-old granddaughter. Jerry was happy when they ordered a burger to share. "Cook it medium," the man said. It looked like the premade burgers would stay on the counter for a while.

Jerry went back to the kitchen and put a fresh patty on the grill. Then he wiped off the cutting board he had used earlier for prepping raw chicken. He sliced the tomatoes and onion. When the burger just passed medium-rare, he plated it up.

When Jerry delivered the food, the little girl asked for a glass of water. Jerry grabbed a glass and used it to scoop up some ice. But the glass broke. Jerry carefully picked the broken glass out of the ice machine. Then he got the girl a fresh glass for the ice water.

① What did Jerry do wrong?

_____

_____

_____

② What could Jerry have done differently?

_____

_____

_____

For answers, please turn to page 1.10.

Study Questions

Circle the best answer to each question below.

1 **Why are elderly people at a higher risk for foodborne illnesses?**

A Their immune systems have weakened with age.

B They are more likely to spend time in a hospital.

C They are more likely to suffer allergic reactions.

D Their appetites have decreased with age.

2 **The three categories of food safety hazards are biological, physical, and**

A temporal.

B practical.

C chemical.

D thermal.

3 **For a foodborne illness to be considered an "outbreak," a minimum of how many people must experience the same illness after eating the same food?**

A 1

B 2

C 10

D 20

4 **The three keys to food safety are practicing good personal hygiene, preventing cross-contamination, and**

A bacteria abatement.

B toxic-metal leaching.

C pathogen measurement.

D time-temperature control.

5 **According to the CDC, the five common causes for foodborne illnesses are failing to cook food adequately, holding food at incorrect temperatures, using contaminated equipment, practicing poor personal hygiene, and**

A reheating leftover food.

B serving ready-to-eat food.

C using single-use, disposable gloves.

D purchasing food from unsafe sources.

For answers, please turn to page 1.10.

Answers

1.4   Who's at Risk?

1, 2, 3, 7, and 8 should be marked.

1.7   What's the Problem?

① Time-temperature abuse
② Poor personal hygiene
③ Time-temperature abuse
④ Poor personal hygiene
⑤ Poor personal hygiene
⑥ Cross-contamination

1.8   Chapter Review Case Study

① Here is what Jerry did wrong.

- He came into work with a stomachache.
- He left cooked burgers sitting out at room temperature. This is time-temperature abuse.
- He didn't wash hands after using the restroom. This is poor personal hygiene.
- He sliced tomatoes on a cutting board that had been used for chicken. This is cross-contamination.
- He scooped ice with a glass. The broken glass in the ice machine is a physical hazard.
- While scooping the ice, Jerry's hand could have touched the ice, leading to cross-contamination of the ice.

② Here is what Jerry could have done differently.

- He should have called in sick. If Jerry was feeling ill, there's a chance he could have made his customers sick.
- When he realized he had made too many hamburger patties, he should have either stored the burgers in hot-holding or thrown them out.
- He should have washed his hands after using the bathroom and after touching his hair.
- When slicing the tomatoes, Jerry should have first washed, rinsed, and sanitized the cutting board. Better yet, he could have used a separate cutting board.
- He should have used an appropriate scoop for the ice.

1.9   Study Questions

① A
② C
③ B
④ D
⑤ D

Notes

2

The Microworld

In the News

Foodborne Illness at a Local Café

Dozens of people became sick with botulism recently at a small café. Customers who ate the restaurant's famous Baked Potato Salad began calling to complain of nausea and vomiting within two days after eating the dish. They eventually also experienced double vision and difficulty in speaking and swallowing.

Health department officials found that the baked potatoes used in the salad were the source of the outbreak. The potatoes had been wrapped in aluminum foil when they were baked. Then they were left on a prep table overnight to cool. Ultimately, the potatoes were left at room temperature for almost 18 hours before they were used in the salad. The bacteria *Clostridium botulinum,* which causes botulism, had the right conditions for growth—time, temperature, and a lack of oxygen provided by the potatoes' foil wrapping.

Working with the inspector, the café changed its procedures for making the salad. Now the employees unwrap and chop the potatoes as soon as they come out of the oven. Then the potatoes are placed on sheet pans in the cooler. Employees monitor and record the potatoes' temperature to make sure they cool in a safe amount of time.

You Can Prevent This

Illnesses from pathogens such as bacteria can be prevented if you understand how pathogens grow and contaminate food. In this chapter, you will learn about the following topics.

- The types of pathogens that cause illness
- What pathogens need to grow
- Food most likely to become unsafe
- The major foodborne illnesses and their characteristics

Concepts from Earlier Chapters

Before reading this chapter, remember these concepts and facts.

Pathogens   Certain viruses, bacteria, parasites, and fungi that can cause illness.

Time-temperature abuse   Food has been time-temperature abused when it stays too long at temperatures that are good for pathogen growth.

Foodborne illness   Disease transmitted to people by food.

Pathogens

Microorganisms are small, living organisms that can be seen only through a microscope. Many microorganisms are harmless, but some can cause illness. Harmful microorganisms are called pathogens. Some pathogens can make you sick when you eat them. Others produce poisons—or toxins—that make you sick. Understanding pathogens is the first step to preventing foodborne-illness outbreaks.

Types of Pathogens

There are four types of pathogens that can contaminate food and cause foodborne illness.

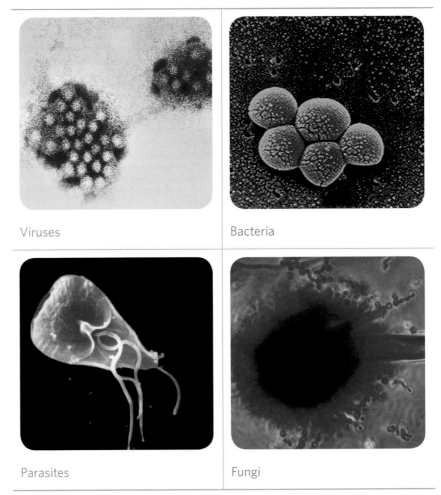

Viruses

Bacteria

Parasites

Fungi

Photos courtesy of the Centers for Disease Control and Prevention

Many viruses, bacteria, and parasites make people sick, but they cannot be seen, smelled, or tasted. On the other hand, some fungi, like mold, change the appearance, smell, or taste of food, but they may not cause illness.

What Pathogens Need to Grow

Understanding how pathogens grow can help you prevent foodborne-illness outbreaks. Pathogens need six conditions to grow. You can remember these conditions by thinking of the words FAT TOM.

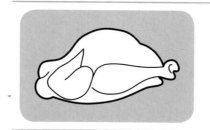

Food

To grow, pathogens need an energy source, such as carbohydrates or proteins.

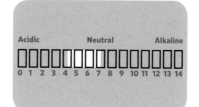

Acidity

Pathogens grow best in food that contains little or no acid.

Temperature

Pathogens grow well in food held between the temperatures of 41°F and 135°F (5°C and 57°C). This range is known as the temperature danger zone.

Time

Pathogens need time to grow. When food is in the temperature danger zone, pathogens grow. After four hours, they will grow to a level high enough to make someone sick.

Oxygen

Some pathogens need oxygen to grow. Others grow when oxygen is not there.

Moisture

Pathogens need moisture in food to grow.

Food Most Likely to Become Unsafe

Any type of food can be contaminated. But some types are better for the growth of pathogens.

- Milk and dairy products

- Eggs (except those treated to eliminate *Salmonella* spp.)

- Meat: beef, pork, and lamb

- Poultry

- Fish

- Shellfish and crustaceans

- Baked potatoes

- Heat-treated plant food, such as cooked rice, beans, and vegetables

- Tofu or other soy protein
- Synthetic ingredients, such as textured soy protein in meat alternatives

- Sprouts and sprout seeds

- Sliced melons
- Cut tomatoes
- Cut leafy greens (fresh leafy greens that have been cut, shredded, sliced, chopped, or torn)

- Untreated garlic-and-oil mixtures

All these types of food have the right FAT TOM conditions that pathogens need to grow. They have a natural potential for contamination because of the way they are grown, produced, or processed. They are also commonly involved in foodborne-illness outbreaks.

Controlling the Growth of Pathogens

You can help keep food safe by controlling FAT TOM. In your operation, however, you will most likely be able to control only time and temperature. These two conditions are so important that the food listed on the previous page is known as food that needs time and temperature control for safety, or TCS food for short.

To control temperature, you must do your best to keep TCS food out of the temperature danger zone. To control time, you must limit how long the TCS food spends in the temperature danger zone. The prep chef in the photo at left is doing this by refrigerating pasta salad right after prepping it.

Like TCS food, ready-to-eat food also needs careful handling to prevent contamination. Ready-to-eat food is exactly what it sounds like: food that can be eaten without further preparation, washing, or cooking. Here are some examples.

• Washed fruit and vegetables, both whole and cut

• Deli meat

• Bakery items

• Sugar, spices, and seasonings

• Cooked food

Apply Your Knowledge

What I Need to Grow

Write an ✘ next to each condition that supports the growth of pathogens in food.

① _____ High fat content

② _____ Protein

③ _____ High acidity

④ _____ Temperature of 155°F (68°C) or higher

⑤ _____ Dry environment

TCS or Not?

Write an ✘ next to each food item that supports the growth of pathogens.

① _____ Raw carrots

② _____ Sliced melons

③ _____ Flour

④ _____ Soda crackers

⑤ _____ Limes

⑥ _____ Milk

For answers, please turn to page 2.36.

Viruses

Viruses are the leading cause of foodborne illness. As a manager, you must understand what viruses are. You should also know the major foodborne illnesses they can cause. Most important, you must learn how to keep them from making your customers sick.

General Information about Viruses

Viruses share some basic characteristics.

Temperature   Viruses can survive cooler and freezer temperatures.

Growth   Viruses can't grow in food. But once eaten, they grow inside a person's intestines.

Contamination   Viruses can contaminate both food and water.

Transfer   Viruses can be transferred from person to person, from people to food, and from people to food-contact surfaces.

When customers get sick from food contaminated with viruses, it's usually because their food was handled by an employee who has a virus. This might be the operation's employee, an employee of the manufacturer, or anyone who has the virus and then handles the food.

People carry viruses in their feces and can transfer them to their hands after using the restroom. Ready-to-eat food can then become contaminated if hands are not washed the right way. Here are the best ways to prevent the spread of viruses in your operation.

Keep foodhandlers who are vomiting or have diarrhea or jaundice from working.

Make sure foodhandlers wash their hands.

Minimize bare-hand contact with ready-to-eat food.

Major Foodborne Illnesses Caused by Viruses

Hepatitis A and Norovirus gastroenteritis are two major foodborne illnesses caused by viruses. For each illness, you must understand the following characteristics.

* Common source

* Food commonly linked with it

* Most common symptoms

* Most important prevention measures

The table on page 2.8 is an overview of all the illnesses in this section. It will help you see similarities and differences that make it easier to remember each illness.

Throughout this chapter, you will also see that the illnesses have been grouped according to their most important prevention measure. Each illness will be grouped by one of the following measures.

* Controlling time and temperature

* Preventing cross-contamination

* Practicing personal hygiene

* Purchasing from approved, reputable suppliers

Note that this measure is not the only way to prevent each illness. Other measures are listed in the tables for each illness.

Practicing Personal Hygiene

These illnesses can be prevented by practicing personal hygiene.

* Hepatitis A   **page 2.9**

* Norovirus gastroenteritis   **page 2.9**

Major Foodborne Illnesses Caused by Viruses

Most Important Prevention Measure			Controlling time and temperature	Preventing cross-contamination	Practicing personal hygiene		Purchasing from approved, reputable suppliers
Illness					Hepatitis A	Norovirus gastroenteritis	
Virus Characteristics	Commonly Linked Food	Poultry					
		Eggs					
		Meat					
		Fish					
		Shellfish			•	•	
		Ready-to-eat food			•	•	
		Produce					
		Rice/grains					
		Milk/dairy products					
		Contaminated water			•	•	
	Most Common Symptoms	Diarrhea				•	
		Abdominal pain/cramps			•	•	
		Nausea			•	•	
		Vomiting				•	
		Fever			•		
		Headache					
	Prevention Measures	Handwashing			•	•	
		Cooking					
		Holding					
		Cooling					
		Reheating					
		Approved suppliers			•	•	
		Excluding foodhandlers			•	•	
		Preventing cross-contamination					

Most Important Prevention Measure: Practicing personal hygiene

Illness	Hepatitis A *(HEP-a-TI-tiss)*
Virus	Hepatitis A

Hepatitis A is mainly found in the feces of people infected with it. The virus can contaminate water and many types of food. It is commonly linked with ready-to-eat food. However, it has also been linked with shellfish contaminated by sewage.

The virus is often transferred to food when infected foodhandlers touch food or equipment with fingers that have feces on them. Eating only a small amount of the virus can make a person sick. An infected person may not show symptoms for weeks but can be very infectious. Cooking does not destroy hepatitis A.

Food Commonly Linked with the Virus

- Ready-to-eat food
- Shellfish from contaminated water

Most Common Symptoms

- Fever (mild)
- General weakness
- Nausea
- Abdominal pain
- Jaundice (appears later)

Other Prevention Measures

- Keep employees who have jaundice out of the operation.
- Keep employees who have been diagnosed with hepatitis A out of the operation.
- Wash hands.
- Minimize bare-hand contact with ready-to-eat food.
- Purchase shellfish from approved, reputable suppliers.

Most Important Prevention Measure: Practicing personal hygiene

Illness	Norovirus gastroenteritis *(NOR-o-VI-rus GAS-tro-EN-ter-I-tiss)*
Virus	Norovirus

Like hepatitis A, Norovirus is commonly linked with ready-to-eat food. It has also been linked with contaminated water. Norovirus is often transferred to food when infected foodhandlers touch food or equipment with fingers that have feces on them.

Eating only a small amount of Norovirus can make a person sick. It is also very contagious. People become contagious within a few hours after eating it. The virus is often in a person's feces for days after symptoms have ended.

Food Commonly Linked with the Virus

- Ready-to-eat food
- Shellfish from contaminated water

Most Common Symptoms

- Vomiting
- Diarrhea
- Nausea
- Abdominal cramps

Other Prevention Measures

- Keep employees with diarrhea and vomiting out of the operation.
- Keep employees who have been diagnosed with Norovirus out of the operation.
- Wash hands.
- Minimize bare-hand contact with ready-to-eat food.
- Purchase shellfish from approved, reputable suppliers.

Apply Your Knowledge

Who Am I?

Identify the virus from the description given and write its name in the space provided.

① _____

- I am commonly linked with ready-to-eat food.
- I am found in the feces of infected people.
- I can produce a mild fever and general weakness.
- Purchasing shellfish from an approved, reputable supplier is one way to prevent me.

② _____

- I am commonly linked with ready-to-eat food.
- I am found in the feces of infected people.
- Handwashing can prevent me.
- Keeping foodhandlers with vomiting or diarrhea out of the operation is one way to control me.

For answers, please turn to page 2.36.
For answers, please turn to page 2.36.

Bacteria

Most foodborne illnesses are caused by viruses. But bacteria can also make people sick. Knowing what bacteria are and how they grow can help you control them.

Characteristics of Bacteria That Cause Foodborne Illness

Bacteria that cause foodborne illness have some basic characteristics.

Temperature   Most bacteria are controlled by keeping food out of the temperature danger zone.

Growth   If FAT TOM conditions are right, bacteria will grow rapidly, as shown at left.

Form   Some bacteria change into a different form, called spores, to keep from dying when they don't have enough food. They can change back and grow again when the food they are on has been time-temperature abused.

Toxin production   Some bacteria make toxins in food as they grow and die. People who eat the toxins can become sick. Cooking may not destroy these toxins.

Major Foodborne Illnesses Caused by Bacteria

For each major foodborne illness caused by bacteria, you must understand these characteristics.

- Common source
- Food commonly linked with it
- Most common symptoms
- Most important prevention measures

The table on page 2.13 is an overview of all the illnesses in this section. It will help you see similarities and differences that make it easier to remember each illness.

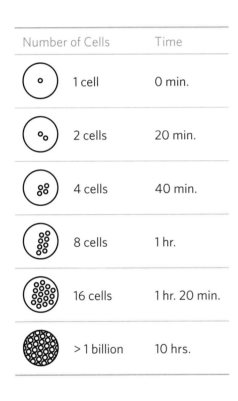

Number of Cells	Time
1 cell	0 min.
2 cells	20 min.
4 cells	40 min.
8 cells	1 hr.
16 cells	1 hr. 20 min.
> 1 billion	10 hrs.

Controlling Time and Temperature

These illnesses can be prevented through time and temperature control.

- *Bacillus cereus* gastroenteritis   **page 2.14**
- Listeriosis   **page 2.14**
- Hemorrhagic colitis   **page 2.15**
- *Clostridium perfringens* gastroenteritis   **page 2.15**
- Botulism   **page 2.16**

Preventing Cross-Contamination

These illnesses can be prevented by preventing cross-contamination.

- Salmonellosis   **page 2.16**

Practicing Personal Hygiene

These illnesses can be prevented by practicing personal hygiene.

- Shigellosis   **page 2.17**
- Staphylococcal gastroenteritis   **page 2.17**

Purchasing from Approved, Reputable Suppliers

These illnesses can be prevented by purchasing products from approved, reputable suppliers.

- *Vibrio vulnificus* primary septicemia/gastroenteritis   **page 2.18**

Major Foodborne Illnesses Caused by Bacteria

Most Important Prevention Measure

			Controlling time and temperature					Preventing cross-contamination	Practicing personal hygiene		Purchasing from approved, reputable suppliers
Illness											
Bacteria Characteristics	Commonly Linked Food		*Bacillus cereus* gastroenteritis	Listeriosis	Hemorrhagic colitis	*Clostridium perfringens* gastroenteritis	Botulism	Salmonellosis	Shigellosis	Staphylococcal gastroenteritis	*Vibrio vulnificus* primary septicemia/gastroenteritis
		Poultry				•		•			
		Eggs						•			
		Meat	•	•	•	•					
		Fish									
		Shellfish									•
		Ready-to-eat food		•					•	•	
		Produce	•		•		•	•	•		
		Rice/grains	•								
		Milk/dairy products	•	•				•			
		Contaminated water							•		•
	Most Common Symptoms	Diarrhea	•		•	•		•	•		•
		Abdominal pain/cramps			•	•		•	•	•	
		Nausea	•				•			•	
		Vomiting	•				•	•		•	•
		Fever						•	•		•
		Headache									
	Prevention Measures	Handwashing							•	•	
		Cooking	•	•	•			•			•
		Holding	•			•	•			•	
		Cooling	•			•	•			•	
		Reheating				•	•			•	
		Approved suppliers			•						•
		Excluding foodhandlers			•			•	•		
		Preventing cross-contamination		•	•			•			

Most Important Prevention Measure: Controlling time and temperature

---

Illness *Bacillus cereus* gastroenteritis *(ba-SIL-us SEER-ee-us GAS-tro-EN-ter-I-tiss)*

Bacteria *Bacillus cereus*

Bacillus cereus is a spore-forming bacteria found in soil. The bacteria can produce two different toxins when allowed to grow to high levels. The toxins cause different illnesses.

Food Commonly Linked with the Bacteria

Diarrhea illness

- Cooked vegetables
- Meat products
- Milk

Vomiting illness

- Cooked rice dishes, including fried rice and rice pudding

Most Common Symptoms

Diarrhea illness

- Watery diarrhea
- No vomiting

Vomiting illness

- Nausea
- Vomiting

Other Prevention Measures

- Cook food to minimum internal temperatures.
- Hold food at the right temperatures.
- Cool food correctly.

Most Important Prevention Measure: Controlling time and temperature

---

Illness Listeriosis *(liss-TEER-ee-O-sis)*

Bacteria *Listeria monocytogenes* *(liss-TEER-ee-uh MON-o-SI-TAHJ-uh-neez)*

Listeria monocytogenes is found in soil, water, and plants. Unlike other bacteria, it grows in cool, moist environments. The illness is uncommon in healthy people, but high-risk populations are especially vulnerable—particularly pregnant women.

Food Commonly Linked with the Bacteria

- Raw meat
- Unpasteurized dairy products
- Ready-to-eat food, such as deli meat, hot dogs, and soft cheeses

Most Common Symptoms

Pregnant women

- Miscarriage

Newborns

- Sepsis
- Pneumonia
- Meningitis

Other Prevention Measures

- Throw out any product that has passed its use-by or expiration date.
- Cook raw meat to minimum internal temperatures.
- Prevent cross-contamination between raw or undercooked food and ready-to-eat food.
- Avoid using unpasteurized dairy products

Illness	**Hemorrhagic colitis** *(hem-or-RA-jik ko-LI-tiss)*
Bacteria	**Shiga toxin-producing** *Escherichia coli* *(ess-chur-EE-kee-UH KO-LI)*, **including O157:H7, O26:H11, O111:H8, and O158:NM**

Shiga toxin-producing *E. coli* can be found in the intestines of cattle. It can contaminate meat during slaughtering. Eating only a small amount of shiga toxin-producing *E. coli* can make a person sick. Once eaten, it produces toxins in the intestines, which cause the illness. The bacteria are often in a person's feces for weeks after symptoms have ended.

Food Commonly Linked with the Bacteria

- Ground beef (raw and undercooked)
- Contaminated produce

Most Common Symptoms

- Diarrhea (eventually becomes bloody)
- Abdominal cramps
- Kidney failure (in severe cases)

Other Prevention Measures

- Cook food, especially ground beef, to minimum internal temperatures.
- Purchase produce from approved, reputable suppliers.
- Prevent cross-contamination between raw meat and ready-to-eat food.
- Keep employees with diarrhea out of the operation.
- Keep employees who have been diagnosed with hemorrhagic colitis out of the operation.

Illness	*Clostridium perfringens* **gastroenteritis** *(klos-TRID-ee-um per-FRIN-jins GAS-tro-EN-ter-I-tiss)*
Bacteria	*Clostridium perfringens*

Clostridium perfringens is found in soil, where it forms spores that allow it to survive. It is also carried in the intestines of both animals and humans.

Clostridium perfringens does not grow at refrigeration temperatures, but it grows very rapidly in food in the temperature danger zone. Commercially prepared food is not often involved in outbreaks. People who get sick usually do not have nausea, fever, or vomiting.

Food Commonly Linked with the Bacteria

- Meat
- Poultry
- Dishes made with meat and poultry, such as stews and gravies

Most Common Symptoms

- Diarrhea
- Severe abdominal pain

Other Prevention Measures

- Cool and reheat food correctly.
- Hold food at the right temperatures.

Most Important Prevention Measure: Controlling time and temperature

Illness   **Botulism** *(BOT-chew-liz-um)*

Bacteria   ***Clostridium botulinum*** *(klos-TRID-ee-um BOT-chew-LINE-um)*

Clostridium botulinum forms spores that are commonly found in water and soil. These spores can contaminate almost any food. The bacteria do not grow well in refrigerated or highly acidic food or in food with low moisture. However, *Clostridium botulinum* grows without oxygen and can produce a deadly toxin when food is time-temperature abused. Without medical treatment, death is likely.

Food Commonly Linked with the Bacteria

- Incorrectly canned food
- Reduced oxygen packaged (ROP) food
- Temperature-abused vegetables, such as baked potatoes
- Untreated garlic-and-oil mixtures

Most Common Symptoms

Initially

- Nausea and vomiting

Later

- Weakness
- Double vision
- Difficulty in speaking and swallowing

Other Prevention Measures

- Hold, cool, and reheat food correctly.
- Inspect canned food for damage.

Most Important Prevention Measure: Preventing cross-contamination

Illness   **Salmonellosis** *(SAL-men-uh-LO-sis)*

Bacteria   ***Salmonella* spp.** *(SAL-me-NEL-uh)*

Many farm animals carry *Salmonella* spp. naturally. Eating only a small amount of these bacteria can make a person sick. How severe symptoms are depends on the health of the person and the amount of bacteria eaten. The bacteria are often in a person's feces for weeks after symptoms have ended.

Food Commonly Linked with the Bacteria

- Poultry and eggs
- Dairy products
- Produce

Most Common Symptoms

- Diarrhea
- Abdominal cramps
- Vomiting
- Fever

Other Prevention Measures

- Cook poultry and eggs to minimum internal temperatures.
- Prevent cross-contamination between poultry and ready-to-eat food.
- Keep foodhandlers who have been diagnosed with salmonellosis out of the operation.

Most Important Prevention Measure: Practicing personal hygiene

Illness	Shigellosis *(SHIG-uh-LO-sis)*
Bacteria	*Shigella* spp. *(shi-GEL-uh)*

Shigella spp. is found in the feces of humans with shigellosis. Most illnesses occur when people eat contaminated food or water. Flies can also transfer the bacteria from feces to food. Eating only a small amount of these bacteria can make a person sick. High levels of the bacteria are often in a person's feces for weeks after symptoms have ended.

Food Commonly Linked with the Bacteria

- Food that is easily contaminated by hands, such as salads containing TCS food (potato, tuna, shrimp, macaroni, and chicken)
- Food that has made contact with contaminated water, such as produce

Most Common Symptoms

- Bloody diarrhea
- Abdominal pain and cramps
- Fever (occasionally)

Other Prevention Measures

- Keep foodhandlers who have diarrhea out of the operation.
- Keep foodhandlers who have been diagnosed with shigellosis out of the operation.
- Wash hands.
- Control flies inside and outside the operation.

Most Important Prevention Measure: Practicing personal hygiene

Illness	Staphylococcal gastroenteritis *(STAF-ul-lo-KOK-al GAS-tro-EN-ter-I-tiss)*
Bacteria	*Staphylococcus aureus* *(STAF-uh-lo-KOK-us OR-ee-us)*

Staphylococcus aureus can be found in humans—particularly in the hair, nose, throat, and infected cuts. It is often transferred to food when people carrying it touch these areas on their bodies and then handle food without washing their hands. If allowed to grow to large numbers in food, the bacteria can produce toxins that cause the illness when eaten. Because cooking cannot destroy these toxins, preventing bacterial growth is critical.

Food Commonly Linked with the Bacteria

Food that requires handling during preparation, including:

- Salads containing TCS food (egg, tuna, chicken, and macaroni)
- Deli meat

Most Common Symptoms

- Nausea
- Vomiting and retching
- Abdominal cramps

Other Prevention Measures

- Wash hands, particularly after touching the hair, face, or body.
- Cover wounds on hands and arms.
- Hold, cool, and reheat food correctly.

Most Important Prevention Measure: Purchasing from approved, reputable suppliers

Illnesses	***Vibrio* gastroenteritis** (VIB-ree-o GAS-tro-EN-ter-I-tiss) ***Vibrio vulnificus* primary septicemia** (VIB-ree-o vul-NIF-ih-kus SEP-ti-SEE-mee-uh)
Bacteria	***Vibrio vulnificus* and *Vibrio parahaemolyticus*** (VIB-ree-o PAIR-uh-HEE-mo-lit-ih-kus)

These bacteria are found in the waters where shellfish are harvested. They can grow very rapidly at temperatures in the middle of the temperature danger zone. People with chronic illnesses (such as diabetes or cirrhosis) who become sick from these bacteria may get primary septicemia, a severe illness that can lead to death.

Food Commonly Linked with the Bacteria

- Oysters from contaminated water

Most Common Symptoms

- Diarrhea
- Abdominal cramps and nausea
- Vomiting
- Low-grade fever and chills

Other Prevention Measures

- Cook oysters to minimum internal temperatures.

Apply Your Knowledge

Who Am I?

Identify the bacteria from the description given and write its name in the space provided.

① _____

- Many farm animals carry me naturally.
- I have been found in eggs, produce, and poultry.
- I can produce diarrhea and vomiting.
- Preventing cross-contamination is one way to prevent me.

② _____

- I can produce toxins if I grow to large numbers.
- I have been linked with salads containing TCS food.
- I can produce retching and abdominal cramps.
- Washing hands can prevent me.

③ _____

- I am found in soil.
- I have been linked with cooked rice dishes.
- I can produce watery diarrhea.
- Cooking, holding, and cooling food correctly can prevent me.

④ _____

- I form spores.
- I have been linked with meat and poultry.
- I do not typically produce fever or vomiting.
- Holding, cooling, and reheating food correctly can prevent me.

⑤ _____

- I do not need oxygen to grow.
- I have been linked with canned and ROP food.
- I can produce double vision.
- Inspecting canned food for damage can prevent me.

⑥ _____

- I am found in the feces of the people I make sick.
- I can cause bloody diarrhea and fever.
- I can be in feces for weeks after symptoms have ended.
- Washing hands can prevent me.

For answers, please turn to page 2.36.

Parasites

Illnesses from parasites are not as common as those caused by bacteria or viruses. But it is still important to understand this group of pathogens so you can prevent the illnesses they cause.

Characteristics of Parasites

Parasites share some common characteristics.

Growth   Parasites cannot grow in food. They need to be in the meat of another animal to survive.

Transfer   Eating food contaminated with parasites will cause illness. Many animals can be hosts. Examples include cows, chickens, pigs, and fish. Parasites can also be found in the feces of animals and people.

Contamination   Parasites can contaminate both food and water—particularly water used to irrigate produce.

Major Foodborne Illnesses Caused by Parasites

For each major foodborne illness caused by parasites, you must understand these characteristics.

* Common source

* Food commonly linked with it

* Most common symptoms

* Most important prevention measures

The table on page 2.20 is an overview of all the illnesses in this section. It will help you see similarities and differences that make it easier to remember each illness.

Purchasing from Approved, Reputable Suppliers

These illnesses can be prevented by purchasing products from approved, reputable suppliers.

* Anisakiasis   **page 2.21**

* Cryptosporidiosis   **page 2.21**

* Giardiasis   **page 2.22**

Major Foodborne Illnesses Caused by Parasites

Most Important Prevention Measure			Controlling time and temperature	Preventing cross-contamination	Practicing personal hygiene	Purchasing from approved, reputable suppliers		
Illness						Anisakiasis	Cryptosporidiosis	Giardiasis
Parasite Characteristics	Commonly Linked Food	Poultry						
		Eggs						
		Meat						
		Fish				•		
		Shellfish						
		Ready-to-eat food						
		Produce					•	•
		Rice/grains						
		Milk/dairy products						
		Contaminated water					•	•
	Most Common Symptoms	Diarrhea					•	•
		Abdominal pain/cramps					•	•
		Nausea					•	•
		Vomiting						
		Fever						•
		Headache						
	Prevention Measures	Handwashing					•	•
		Cooking				•		
		Holding						
		Cooling						
		Reheating						
		Approved suppliers				•	•	•
		Excluding foodhandlers					•	•
		Preventing cross-contamination						

Most Important Prevention Measure: Purchasing from approved, reputable suppliers

| **Illness** | **Anisakiasis** *(ANN-ih-SAH-KYE-ah-sis)* |
| **Parasite** | ***Anisakis simplex*** *(ANN-ih-SAHK-iss SIM-plex)* |

People can get sick when they eat raw or undercooked fish containing this parasite.

Food Commonly Linked with the Parasite

Raw and undercooked fish, including:

- Herring
- Cod
- Halibut
- Mackerel
- Pacific salmon

Most Common Symptoms

- Tingling in throat
- Coughing up worms

Other Prevention Measures

- Cook fish to minimum internal temperatures.
- If serving raw or undercooked fish, purchase sushi-grade fish that has been frozen to the right time-temperature requirements.

Most Important Prevention Measure: Purchasing from approved, reputable suppliers

| **Illness** | **Cryptosporidiosis** *(KRIP-TOH-spor-id-ee-O-sis)* |
| **Parasite** | ***Cryptosporidium parvum*** *(KRIP-TOH-spor-ID-ee-um PAR-vum)* |

Cryptosporidium parvum can be found in the feces of people infected with it. Foodhandlers can transfer it to food when they touch food with fingers that have feces on them. Day-care and medical communities have been frequent locations of person-to-person spread of this parasite. Symptoms will be more severe in people with weakened immune systems.

Food Commonly Linked with the Parasite

- Contaminated water
- Produce

Most Common Symptoms

- Watery diarrhea
- Abdominal cramps
- Nausea
- Weight loss

Other Prevention Measures

- Use properly treated water.
- Keep foodhandlers with diarrhea out of the operation.
- Wash hands.

Photo courtesy of Boskovich Farms, Inc.

Most Important Prevention Measure: Purchasing from approved, reputable suppliers

Illness	**Giardiasis** (*JEE-are-DYE-uh-sis*)
Parasite	***Giardia duodenalis*** (*jee-ARE-dee-uh do-WAH-den-AL-is*), also known as ***G. lamblia*** or ***G. intestinalis***

Giardia duodenalis can be found in the feces of infected people. Foodhandlers can transfer the parasite to food when they touch food with fingers that have feces on them.

Food Commonly Linked with the Parasite

- Improperly treated water
- Produce

Most Common Symptoms

Initially

- Fever

Later

- Diarrhea
- Abdominal cramps
- Nausea

Other Prevention Measures

- Use properly treated water.
- Keep foodhandlers with diarrhea out of the operation.
- Wash hands.

Photo courtesy of Boskovich Farms, Inc.

Apply Your Knowledge

Who Am I?

Identify the parasite from the description given and write its name in the space provided.

① _____

- I have been found in produce.
- I have been found in contaminated water.
- I can produce a fever and diarrhea.
- Excluding foodhandlers with diarrhea from the operation is one way to prevent me.

② _____

- Cooking fish can destroy me.
- I have been found in mackerel.
- People infected with me cough up worms.
- Purchasing fish from approved, reputable suppliers is one way to prevent me.

③ _____

- I can be spread by fingers with feces on them.
- I have been found in contaminated produce.
- I can produce watery diarrhea and weight loss.
- Purchasing produce from approved, reputable suppliers is one way to prevent me.

For answers, please turn to page 2.36.

Fungi

So far, you have learned about pathogens that cause foodborne illness. Fungi are pathogens that only sometimes make people sick. Mostly they spoil food. They are found in air, soil, plants, water, and some food. Mold (as shown in the photo of green beans at left) and yeast are examples.

Molds

Molds share some basic characteristics.

Effects   Molds spoil food and sometimes cause illness.

Toxins   Some molds produce toxins, such as aflatoxins.

Growth   Molds grow under almost any condition. But they grow well in acidic food with little moisture. Examples are jams, jellies, and cured, salty meat such as ham, bacon, and salami.

Temperature   Cooler or freezer temperatures may slow the growth of molds, but they don't kill them.

Prevention measure   Throw out all moldy food, unless the mold is a natural part of the product (e.g., cheese such as Brie, Camembert, and Gorgonzola, as shown in the photo at left). The Food and Drug Administration (FDA) recommends cutting away moldy areas in hard cheese—at least one inch (2.5 centimeters) around them. You can also use this procedure on food such as salami and firm fruit and vegetables.

Yeasts

Yeasts share some basic characteristics.

Signs of spoilage   Yeasts can spoil food quickly. Signs of spoilage can include a smell or taste of alcohol. The yeast itself may look like a white or pink discoloration or slime, as shown in the photo of jelly at left. It also may bubble.

Growth   Like molds, yeasts grow well in acidic food with little moisture, such as jellies, jams, syrup, honey, and fruit or fruit juice.

Prevention measure   Throw out any food that has been spoiled by yeast.

Apply Your Knowledge

Who Am I?

Identify the fungi from the description given and write its name in the space provided.

① _____

- I produce an alcohol-like odor.
- I may look pink or slimy.
- I may cause food to bubble.
- Food containing me should be thrown out.

② _____

- Some forms of me can produce a dangerous toxin.
- I am not killed by freezer temperatures.
- I grow well in food such as jelly and salami.
- Sometimes I can be cut away from food.

For answers, please turn to page 2.36.

Biological Toxins

You learned earlier in this chapter that most foodborne illnesses are caused by pathogens, a form of biological contamination. But you also must be aware of biological toxins that can make people sick. Biological toxins are made by pathogens, or they come from a plant or an animal.

Seafood toxins, plant toxins, and mushroom toxins can all cause foodborne illnesses. So you must understand what these toxins are and the illnesses they can cause. Most important, you must learn how to keep them from making your customers sick.

Seafood Toxins

Seafood toxins can't be smelled or tasted. They also can't be destroyed by freezing or cooking once they form in food. Below are the two groups of seafood toxins.

Fish toxins   Some fish toxins are a natural part of the fish. Others are made by pathogens on it. Some fish become contaminated when they eat smaller fish that have eaten a toxin.

Shellfish toxins   Shellfish, such as the oysters in the photo at left, can be contaminated when they eat marine algae that have a toxin.

Major Foodborne Illnesses Caused by Seafood Toxins

For each major foodborne illness caused by seafood toxins, you must understand these characteristics.

- Common source
- Food commonly linked with it
- Most common symptoms
- Most important prevention measures

The table on page 2.26 is an overview of all the illnesses in this section. It will help you see similarities and differences that make it easier to remember each illness.

Purchasing from Approved, Reputable Suppliers

These illnesses can be prevented by purchasing products from approved, reputable suppliers.

- Scombroid poisoning   **page 2.27**
- Ciguatera fish poisoning   **page 2.27**
- Paralytic shellfish poisoning (PSP)   **page 2.28**
- Neurotoxic shellfish poisoning (NSP)   **page 2.28**
- Amnesic shellfish poisoning (ASP)   **page 2.29**

Major Foodborne Illnesses Caused by Seafood Toxins

Most Important Prevention Measure			Controlling time and temperature	Preventing cross-contamination	Practicing personal hygiene	Purchasing from approved, reputable suppliers				
Illness						Scombroid poisoning	Ciguatera fish poisoning	Paralytic shellfish poisoning (PSP)	Neurotoxic shellfish poisoning (NSP)	Amnesic shellfish poisoning (ASP)
Seafood Toxin Characteristics	Commonly Linked Food	Fish				•	•			
		Shellfish						•	•	•
	Most Common Symptoms	Diarrhea				•		•	•	•
		Abdominal pain/cramps								•
		Nausea					•	•		
		Vomiting				•	•	•	•	•
		Fever								
		Headache				•				
		Neurological symptoms				•	•	•	•	•
	Prevention Measures	Handwashing								
		Cooking								
		Holding				•				
		Cooling								
		Reheating								
		Approved suppliers				•	•	•	•	•
		Excluding foodhandlers								
		Preventing cross-contamination								

Most Important Prevention Measure: Purchasing from approved, reputable suppliers

Illness	**Scombroid poisoning** (SKOM-broyd)
Toxin	**Histamine** (HISS-ta-meen)

Scombroid poisoning is also known as histamine poisoning. It is an illness caused by eating high levels of histamine in scombroid and other species of fish. When the fish are time-temperature abused, bacteria on the fish make the toxin. It cannot be destroyed by freezing, cooking, smoking, or curing.

Food Commonly Linked with the Toxin

- Tuna
- Bonito
- Mackerel
- Mahi mahi

Most Common Symptoms

Initially

- Reddening of the face and neck
- Sweating
- Headache
- Burning or tingling sensation in the mouth or throat

Possibly later

- Diarrhea
- Vomiting

Other Prevention Measures

- Prevent time-temperature abuse during storage and preparation.

Most Important Prevention Measure: Purchasing from approved, reputable suppliers

Illness	**Ciguatera fish poisoning** (SIG-wa-TAIR-uh)
Toxin	**Ciguatoxin** (SIG-wa-TOX-in)

Ciguatoxin is found in certain marine algae. The toxin builds up in certain fish when they eat smaller fish that have eaten the toxic algae. Ciguatoxin cannot be detected by smell or taste. Cooking or freezing the fish will not eliminate it. Symptoms may last months or years depending on how severe the illness is.

Food Commonly Linked with the Toxin

Predatory tropical reef fish from the Pacific Ocean, the western part of the Indian Ocean, and the Caribbean Sea, including:

- Barracuda
- Grouper
- Jacks
- Snapper

Most Common Symptoms

- Reversal of hot and cold sensations
- Nausea
- Vomiting
- Tingling in fingers, lips, or toes
- Joint and muscle pain

Other Prevention Measures

- Purchase predatory tropical reef fish from approved, reputable suppliers.

Illness	**Paralytic shellfish poisoning (PSP)** *(PAIR-ah-LIT-ik)*
Toxin	**Saxitoxin** *(SAX-ih-TOX-in)*

Some types of shellfish can become contaminated as they filter toxic algae from the water. People get sick with paralytic shellfish poisoning (PSP) when they eat these shellfish. Saxitoxin cannot be smelled or tasted. It is not destroyed by cooking or freezing. Death from paralysis may result if high levels of the toxin are eaten.

Food Commonly Linked with the Toxin

Shellfish found in colder waters, such as those of the Pacific and New England coasts, including:

- Clams
- Mussels
- Oysters
- Scallops

Most Common Symptoms

- Numbness
- Tingling of the mouth, face, arms, and legs
- Dizziness
- Nausea
- Vomiting
- Diarrhea

Other Prevention Measures

- Purchase shellfish from approved, reputable suppliers.

Illness	**Neurotoxic shellfish poisoning (NSP)** *(NUR-o-TOX-ik)*
Toxin	**Brevetoxin** *(BREV-ih-TOX-in)*

Some types of shellfish can become contaminated as they filter toxic algae from the water. People get sick with neurotoxic shellfish poisoning (NSP) when they eat these shellfish. Brevetoxin cannot be smelled or tasted. It is not destroyed by cooking or freezing.

Food Commonly Linked with the Toxin

Shellfish found in the warmer waters of the west coast of Florida, the Gulf of Mexico, and the Caribbean Sea, including:

- Clams
- Mussels
- Oysters

Most Common Symptoms

- Tingling and numbness of the lips, tongue, and throat
- Dizziness
- Reversal of hot and cold sensations
- Vomiting
- Diarrhea

Other Prevention Measures

- Purchase shellfish from approved, reputable suppliers.

Most Important Prevention Measure: Purchasing from approved, reputable suppliers

Illness	Amnesic shellfish poisoning (ASP) *(am-NEE-zik)*
Toxin	Domoic acid *(duh-MO-ik)*

Some types of shellfish can become contaminated as they filter toxic algae from the water. People get sick with amnesic shellfish poisoning (ASP) when they eat these shellfish. The severity of symptoms depends on the amount of toxin eaten and the health of the person. Domoic acid cannot be smelled or tasted. It is not destroyed by cooking or freezing.

Food Commonly Linked with the Toxin

Shellfish found in the coastal waters of the Pacific Northwest and the east coast of Canada, including:

- Clams
- Mussels
- Oysters
- Scallops

Most Common Symptoms

Initially

- Vomiting
- Diarrhea
- Abdominal pain

Possibly later

- Confusion
- Memory loss
- Disorientation
- Seizure
- Coma

Other Prevention Measures

- Purchase shellfish from approved, reputable suppliers.

Mushroom Toxins

Foodborne illnesses linked with mushrooms are almost always caused by eating toxic, wild mushrooms collected by amateur hunters. Most cases happen because toxic mushrooms are mistaken for edible ones. The symptoms of illness depend on the type of toxic mushrooms eaten.

Mushroom toxins are not destroyed by cooking or freezing. Do NOT use mushrooms or mushroom products unless you have purchased them from approved, reputable suppliers. A package of approved mushrooms is shown in the photo at left.

Plant Toxins

Plant toxins are another form of biological contamination. Illnesses from plant toxins usually happen because an operation has purchased plants from an unapproved supplier. Some illnesses, however, are caused by plants that haven't been cooked the right way. Here are some examples of items that can make people sick.

- Toxic plants, such as fool's parsley or wild turnips, mistaken for the edible version
- Honey from bees allowed to harvest nectar from toxic plants
- Undercooked kidney beans

Purchase plants and items made with plants only from approved, reputable suppliers. Then cook and hold dishes made from these items correctly.

Apply Your Knowledge

Who Am I?

Identify the seafood toxin from the description given and write its name in the space provided.

① _____

- I build up in predatory tropical reef fish.
- I have been linked with barracuda and snapper.
- I can produce tingling lips and joint pain.
- I cause an illness with symptoms that can last for years.

② _____

- When certain fish are time-temperature abused, bacteria on the fish make me.
- I have been linked with tuna and bonito.
- I can produce a burning sensation in the mouth.
- I cannot be destroyed by smoking or curing.

③ _____

- I am found in toxic marine algae.
- I have been linked with scallops and other seafood.
- I can cause paralysis and death.
- I am found in colder waters, such as those off the New England coast.

④ _____

- I am found in the waters of the Pacific Northwest.
- I can cause memory loss.
- I can produce abdominal pain.
- I have been linked with scallops.

⑤ _____

- I can cause the reversal of the sensations of hot and cold.
- I have been linked with clams.
- I can make lips go numb.
- I can be found in the waters off the west coast of Florida.

For answers, please turn to page 2.36.

Chapter Summary

Microorganisms are small, living organisms. Harmful microorganisms are called pathogens. Understanding how pathogens grow, contaminate food, and affect people will help you understand how to prevent the foodborne-illness outbreaks caused by them. The words FAT TOM will help you remember the conditions that pathogens need to grow: food, acidity, temperature, time, oxygen, and moisture. Any type of food can be contaminated. But some types, known as TCS food, are better for pathogen growth.

The four types of pathogens discussed in this chapter are viruses, bacteria, parasites, and fungi. Viruses are the leading cause of foodborne illness. They cannot grow in food, but they can survive cooler and freezer temperatures. The key to preventing the spread of viruses is good personal hygiene. Bacteria can usually be controlled by keeping food out of the temperature danger zone. Some bacteria can change into spores to keep from dying when they don't have enough food. Others can make toxins in food that will make people who eat the food sick. Parasites need to be in the meat of another animal to survive. They can contaminate both food and water—particularly water used to irrigate produce. Purchasing products from approved, reputable suppliers is important for preventing foodborne illnesses caused by parasites. Fungi only sometimes make people sick. Mostly they spoil food. Examples are mold and yeast. Any food spoiled by mold or yeast should be thrown out, unless the mold is a natural part of the product.

Seafood toxins, plant toxins, and mushroom toxins also can cause foodborne illnesses. Fish toxins can be a natural part of the fish or made by pathogens on it. Some occur when fish eat smaller fish that have the toxin. Shellfish toxins are caused by marine algae that have a toxin, which the shellfish then eats. Foodborne illnesses linked with mushrooms are almost always caused by eating toxic, wild mushrooms collected by amateur hunters. Similarly, foodborne illnesses caused by plant toxins usually happen because an operation has purchased plants from an unapproved food supplier. Purchasing products from approved, reputable suppliers is important for preventing all these types of foodborne illnesses.

Chapter Review Activities

Now take what you have learned in this chapter and apply it to the following activities.

Read each scenario and then answer the question that follows it.

A day-care center decided to prepare stir-fried rice to serve for lunch the next day. The rice was cooked to the right temperature at 1:00 p.m. It was then covered and placed on a countertop, where it was allowed to cool at room temperature. At 6:00 p.m., the cook placed the rice in the refrigerator. At 9:00 a.m. the following day, the rice was combined with the other ingredients for stir-fried rice and cooked to 165°F (74°C) for at least 15 seconds. The cook covered the rice and left it on the stove until noon when she reheated it. Within an hour of eating the rice, several of the children complained that they were nauseous and began to vomit.

① What pathogen caused the illness and why?

_____

_____

Roberto received a shipment of frozen mahi-mahi steaks. The steaks were frozen solid at the time of delivery, and the packages were sealed but contained a large amount of ice crystals, indicating they had been time-temperature abused. Roberto accepted the mahi-mahi steaks and thawed them in the refrigerator at a temperature of 38°F (3°C). The thawed fish steaks were then held at this temperature during the evening shift and were cooked to order. The chefs followed the right guidelines for prepping, cooking, and serving the fish, monitoring time and temperature throughout the process. Unfortunately, the fish steaks caused an outbreak of scombroid poisoning.

② Why did the mahi-mahi steaks cause scombroid poisoning?

_____

_____

Match each foodborne illness with the best method for preventing it.

Prevention Measure

Ⓐ Purchasing from approved, reputable suppliers

Ⓑ Practicing personal hygiene

Ⓒ Controlling time and temperature

Ⓓ Preventing cross-contamination

Foodborne Illness

① _____ Hemorrhagic colitis

② _____ Neurotoxic shellfish poisoning (NSP)

③ _____ Listeriosis

④ _____ Amnesic shellfish poisoning (ASP)

⑤ _____ *Vibrio vulnificus* primary septicemia

⑥ _____ Staphylococcal gastroenteritis

⑦ _____ Norovirus gastroenteritis

⑧ _____ Salmonellosis

⑨ _____ Giardiasis

⑩ _____ Scombroid poisoning

⑪ _____ Shigellosis

⑫ _____ *Bacillus cereus* gastroenteritis

⑬ _____ Botulism

⑭ _____ *Clostridium perfringens* gastroenteritis

⑮ _____ Hepatitis A

⑯ _____ Anisakiasis

⑰ _____ Cryptosporidiosis

⑱ _____ Paralytic shellfish poisoning (PSP)

⑲ _____ Ciguatera fish poisoning

⑳ _____ *Vibrio* gastroenteritis

For answers, please turn to page 2.37.

Study Questions

Circle the best answer to each question below.

1 **Foodborne pathogens grow well at temperatures**

   A below 32°F (0°C).
   B between 1°F to 40°F (–17°C to 4°C).
   C between 41°F to 135°F (5°C to 57°C).
   D above 212°F (100°C).

2 **FAT TOM stands for Food, Acidity, Temperature, Time, Oxygen and**

   A Meat.
   B Moisture.
   C Melatonin.
   D Management.

3 **Which pathogen is primarily found in the hair, nose, and throat of humans?**

   A *Giardia duodenalis*
   B *Bacillus cereus*
   C *Clostridium botulinum*
   D *Staphylococcus aureus*

4 **While commonly linked with contaminated ground beef, what pathogen has also been linked with contaminated produce?**

   A *Bacillus cereus*
   B *Salmonella* spp.
   C Shiga toxin-producing *E. coli*
   D *Clostridium perfringens*

5 **Which practice can reduce *Salmonella* spp. in poultry to safe levels?**

   A Storing food at 55°F (13°C) or higher
   B Inspecting canned food for damage
   C Cooking food to the right temperature
   D Purchasing oysters from approved, reputable suppliers

6 **Covering wounds can help prevent the spread of which pathogen?**

   A *Staphylococcus aureus*
   B Norovirus
   C *Vibrio vulnificus*
   D *Salmonella* spp.

⑦ **Which foodborne illness has been linked with ready-to-eat food and shellfish contaminated by sewage?**

A Hepatitis A

B Anisakiasis

C Shigellosis

D Botulism

⑧ **Viruses such as Norovirus and hepatitis A can be spread when foodhandlers fail to**

A use pasteurized eggs.

B wash their hands.

C determine the correct moisture level.

D purchase beef from approved, reputable suppliers.

⑨ **What is the best way to prevent a foodborne illness caused by seafood toxins?**

A Freezing seafood prior to cooking it

B Purchasing smoked or cured seafood

C Purchasing seafood from approved, reputable suppliers

D Cooking seafood to the right minimum internal temperature

⑩ **A person who ate raw oysters later became disoriented and suffered memory loss. What illness was most likely the cause?**

A Amnesic shellfish poisoning

B Paralytic shellfish poisoning

C Neurotoxic shellfish poisoning

D Hemorrhagic shellfish poisoning

⑪ **Foodservice operations should not use mushrooms unless they have been**

A stored at 41°F (5°C) or lower.

B frozen before cooking or serving.

C purchased from an approved, reputable supplier.

D cooked to an internal temperature of 135°F (57°C).

For answers, please turn to page 2.37.

Answers

2.5   What I Need to Grow

2 should be marked.

2.5   TCS or Not?

2 and 6 should be marked.

2.10   Viruses: Who Am I?

① Hepatitis A

② Norovirus

2.18   Bacteria: Who Am I?

① *Salmonella* spp.

② *Staphylococcus aureus*

③ *Bacillus cereus*

④ *Clostridium perfringens*

⑤ *Clostridium botulinum*

⑥ *Shigella* spp.

2.22   Parasites: Who Am I?

① *Giardia duodenalis*

② *Anisakis simplex*

③ *Cryptosporidium parvum*

2.24   Fungi: Who Am I?

① Yeast

② Mold

2.31   Seafood Toxins: Who Am I?

① Ciguatoxin

② Histamine

③ Saxitoxin

④ Domoic acid

⑤ Brevetoxin

2.32 **Chapter Review Activities**

Scenarios

① *Bacillus cereus* caused the illness. It is commonly linked with cooked rice. The pathogen was allowed to grow when the rice was cooled incorrectly and held at the wrong temperature.

② When the mahi-mahi was time-temperature abused, the bacteria on the fish produced the toxin histamine. Because cooking does not destroy this toxin, eating the fish resulted in scombroid poisoning.

Matching

① C	⑪ B
② A	⑫ C
③ C	⑬ C
④ A	⑭ C
⑤ A	⑮ B
⑥ B	⑯ A
⑦ B	⑰ A
⑧ D	⑱ A
⑨ A	⑲ A
⑩ A	⑳ A

2.34 **Study Questions**

① C

② B

③ D

④ C

⑤ C

⑥ A

⑦ A

⑧ B

⑨ C

⑩ A

⑪ C

3

Contamination and Food Allergens

In the News

Restaurant Group Works to Prevent Allergic Reactions

A Midwestern restaurant group decided to review its food-allergy prevention policies. After looking at its procedures, it developed a system to call out meals for guests with food allergies. According to a spokesperson for the group, the system has been very successful in helping avoid food-allergy problems.

When a guest lets a server know about a food allergy, the server takes the order on a separate, brightly colored ticket. The ticket includes the guest's table number and position at the table, the food allergy with special instructions, and the individual's order. The server then delivers the ticket to the kitchen and confirms it with the chef. The ticket stays with the meal and is signed by the manager before it is served.

You Can Build on This

The operation above took steps to ensure that customers could safely enjoy their food. As a manager, you must make sure your operation's food is safe for your customers. In addition to allergens, contamination can come from chemical and physical sources. In this chapter, you will learn about the following.

- Preventing both accidental and deliberate contamination of food

- Making sure staff knows how to help customers who have food allergies

---

Concepts from Earlier Chapters

Before reading this chapter, remember these concepts and facts.

Chemical hazard   Cleaners, sanitizers, polishes, machine lubricants, and toxic metals that can contaminate food and make people sick.

Physical hazard   Objects, such as hair, dirt, metal staples, bones, and broken glass, that can get into food.

Contamination

You learned in chapter 2 that biological contaminants are the leading cause of foodborne illnesses. But there are other contaminants to watch out for too. Chemicals and physical objects are risks to the food you serve. You also have to take steps to prevent people from deliberately contaminating food.

Chemical Contaminants

Chemicals have caused many cases of foodborne illnesses. These contaminants can come from everyday items found in the operation.

Toxic Metals

Some utensils and equipment contain toxic metals that can contaminate acidic food. A person who then eats this food can get toxic-metal poisoning. This illness is frequently caused by storing or prepping acidic food with equipment containing the following metals.

Lead   This metal is found in pewter, which can be used to make pitchers and other tableware.

Copper   This metal is sometimes found in cookware like pots and pans.

Zinc   This metal is found in galvanized items, which are coated with zinc. Some buckets, tubs, and other items may be galvanized.

To prevent toxic-metal poisoning, you should only use utensils and equipment that are made for handling food.

Foodservice Chemicals

Chemicals can contaminate food if they are used or stored the wrong way. Cleaners, sanitizers, polishes, and machine lubricants pose risks. To keep food safe, follow these guidelines.

- Store chemicals away from food, utensils, and equipment used for food. Keep them in a separate storage area in their original container, as shown in the photo at left.

- Follow the manufacturers' directions when using chemicals.

- Be careful when using chemicals while food is being prepped.

- If you transfer a chemical to a new container, you must label it with the common name of the chemical.

- Only use lubricants that are made for foodservice equipment.

Physical Contaminants

Food can become contaminated when objects get into it. It can also happen when natural objects are left in food, like bones in a fish fillet.

Here are some common physical contaminants.

- Metal shavings from cans, as shown in the photo at left

- Staples from cartons

- Glass from broken light bulbs

- Blades from plastic or rubber scrapers

- Fingernails, hair, and bandages

- Dirt

- Bones

- Jewelry

- Fruit pits, as shown on the spoon in the photo at left

Closely inspect the food you receive. Take steps to make sure no physical contaminants can get into it.

Deliberate Contamination of Food

So far, you have learned about methods to prevent the accidental contamination of food. But you also must take steps to stop people who are actually trying to contaminate it. This may include the following groups.

- Terrorists or activists
- Current or former employees
- Vendors
- Competitors

These people may try to tamper with your food using biological, chemical, or physical contaminants. They may even use radioactive materials. Attacks might occur anywhere in the food supply chain. But attacks are usually focused on a specific food item, process, or business.

The best way to protect food is to make it as hard as possible for someone to tamper with it. For this reason, a food defense program should deal with the points in your operation where food is at risk.

Human element   Anyone who has access to an operation's food can be a security risk. Examples include delivery people, service contractors, guests, and even staff. To lower the risk, you should control access to food and prep areas. For example, make sure only on-duty staff are allowed in work areas. The employee in the photo at left is wearing a name badge. This is one way to make sure that only staff are in work areas. Additionally, you could train employees to report suspicious activities.

Building interior   If areas inside your facility are not secure, your food might not be either. To secure these areas, you should do things like eliminating places for intruders to hide.

Building exterior   Make sure that people can't get inside the facility in unexpected ways. There are several things you can do to prevent this, including lighting the exterior and controlling access to roof vents.

Make sure your employees understand food defense issues. You should start by developing procedures that address each potential threat. Then train employees to follow them.

Apply Your Knowledge

Can It Cause Contamination?

Decide if the situation could cause food to be contaminated. Explain why or why not.

① Jack needed to prep heads of lettuce. When he ripped open the box, two staples flew off. He could only find one of them, which he threw away.

_____

② Lee mixed some sanitizer solution according to the directions on the container. He then used it to sanitize a clean prep table after the kitchen was closed.

_____

③ Anita needed to make several gallons of lemonade for a private outdoor party. She mixed the lemonade and served it in a new, galvanized tub. The tub was cleaned and sanitized before use.

_____

④ Lisa used a new saucepan to make hollandaise sauce. The pan had a copper base but had stainless steel on the inside.

_____

⑤ Pete carried a large tray of dirty dishes to the dishwashing area. A wine glass fell off the tray and broke on a prep table, near some chopped onions. He cleaned up the glass and rinsed the onions in a colander.

_____

⑥ Marlene put a large stock pot on the shelf below the dishwasher detergent.

_____

⑦ Avery put vegetables dressed in vinaigrette on a pewter platter.

_____

⑧ Paul poured some detergent from its original container into a smaller bottle. When he was finished using it, he put both bottles back into the chemical-storage area.

_____

⑨ Jennie wore false fingernails to work. She made potato salad and hamburger patties for most of the morning.

_____

For answers, please turn to page 3.13.

Food Allergens

The number of people in the United States with food allergies is increasing. A food allergy is the body's negative reaction to a food protein. There are specific signs that a customer is having an allergic reaction. To protect your customers, you should be able to recognize these signs and know what to do. You also should know the types of food that most often cause allergic reactions, so you can help prevent a reaction from happening.

Allergy Symptoms

Depending on the person, an allergic reaction can happen right after the food is eaten or several hours later. This reaction could include some or all of these symptoms.

- Itching in and around the mouth, face, or scalp
- Tightening in the throat
- Wheezing or shortness of breath
- Hives, as shown in the photo at left
- Swelling of the face, eyes, hands, or feet
- Abdominal cramps, vomiting, or diarrhea
- Loss of consciousness
- Death

If a customer is having an allergic reaction to food, call the emergency number in your area.

Common Food Allergens

You and your staff must be aware of the most common food allergens and the menu items that contain them.

Milk and dairy products

Eggs and egg products

Fish and shellfish

Wheat

Soy and soy products

Peanuts and tree nuts, such as pecans and walnuts

Preventing Allergic Reactions

Both service staff and kitchen staff need to do their parts to avoid serving food that can cause an allergic reaction.

Service Staff

Your employees should be able to tell customers about menu items that contain potential allergens. At minimum, have one person available per shift to answer customers' questions about menu items. When customers say they have a food allergy, your staff should take it seriously. They must be able to do the following.

Describing dishes   Tell customers how the item is prepared, as the server in the photo at left is doing. Sauces, marinades, and garnishes often contain allergens. For example, peanut butter is sometimes used as a thickener in sauces or marinades. This information is critical to a customer with a peanut allergy.

Identifying ingredients   Identify any "secret" ingredients. For example, your operation may have a house specialty that includes an allergen. While you may not want to share the recipe with the public, staff must be able to tell the secret ingredient to a customer who asks.

Suggesting items   Suggest simple menu items. Complex items such as casseroles, soups, and some desserts may contain many ingredients. These can be difficult to fully describe to customers.

Kitchen Staff

Staff must make sure that allergens are not transferred from food containing an allergen to the food served to the customer. This is called cross-contact. Here are a few examples of how it can happen.

- Cooking different types of food in the same fryer oil can cause cross-contact. In the photo at left, shrimp allergens could be transferred to the chicken being fried in the same oil.

- Putting food on surfaces that have touched allergens can cause cross-contact. For example, putting chocolate chip cookies on the same parchment paper that was used for peanut butter cookies can transfer some of the peanut allergen.

Here is how to avoid cross-contact.

- Wash, rinse, and sanitize cookware, utensils, and equipment before prepping food.

- Wash your hands and change gloves before prepping food.

- Assign specific equipment for prepping food for customers with allergies. For example, if your operation serves fried chicken and fried clams, you could designate one piece of equipment for the seafood and the other for the chicken.

Apply Your Knowledge

Identify the Symptoms

Write an ✘ next to the symptoms that could indicate a customer is having an allergic reaction.

1. _____ Loss of consciousness
2. _____ Bruising
3. _____ Sneezing
4. _____ Coughing
5. _____ Itchy scalp
6. _____ Hives
7. _____ Swollen face
8. _____ Abdominal cramps
9. _____ Swollen abdomen
10. _____ Increased appetite
11. _____ Shortness of breath
12. _____ Tightening in the chest
13. _____ Uncontrollable laughing
14. _____ Tightening in the throat

Identify the Most Common Food Allergens

Write an ✘ next to a food if it is or has a common food allergen.

1. _____ Tea
2. _____ Cod
3. _____ Wheat flour
4. _____ Melons
5. _____ Peanut butter
6. _____ Crab legs
7. _____ Potatoes
8. _____ Mushrooms
9. _____ Tomatoes
10. _____ Pecan pie
11. _____ Citrus fruit
12. _____ Green peppers
13. _____ Squash and eggplant
14. _____ Soybeans
15. _____ Rice and rice products
16. _____ Omelet
17. _____ Vanilla ice cream

Keeping Food Safe for Customers with Allergies

Explain how to handle each situation so the customer's food is safe to eat.

1. Fay is a server at City Star Café. One quiet afternoon, a customer rushed in and let her know he was in a hurry. He also said he was allergic to peanuts. He wanted to order the vegetable egg rolls and asked if they contained any peanut products or were fried in peanut oil. Fay knew the ingredients of most items they served. However, she was not sure about the egg rolls, so she went to the kitchen to ask the chef. The chef was busy receiving a delivery, so Fay asked Larry, the sandwich maker, if he knew. Larry was not sure either. What should Fay do next?

_____

_____

2. Gerard is a cook at The Junction, an independent, quick-service operation. It is known locally for its broasted chicken. Broasting involves cooking the chicken in a pressure cooker with oil. The oil is filtered every night and reused for a limited time. Other popular menu items at The Junction include fried cheese sticks and jalapeno poppers, which are cooked in a separate fryer.

One evening, a server put in an order for two broasted chicken dinners. On the ticket, next to one dinner, he wrote "no dairy." After questioning the server, Gerard learned that the dinner was for a customer who is allergic to dairy. Fortunately, the item does not have any dairy in it. What should Gerard do to make sure the "no dairy" dinner is safe for the allergic customer to eat?

_____

_____

For answers, please turn to page 3.13.

Chapter Summary

Contamination of food can come from many things in your operation. Chemical contaminants include toxic metals, cleaners, sanitizers, polishes, and machine lubricants. To help prevent chemical contamination, use only utensils and equipment that are made for handling food. Also, store them away from utensils and equipment used for food. Always follow the manufacturers' directions when using chemicals.

Physical contamination can happen when objects get into food. Naturally occurring objects, such as bones in a fish fillet, are a physical hazard. Closely inspect the food you receive. Make sure no physical contaminants can get into it at any point during the flow of food.

People may try to tamper with food using biological, chemical, physical, or even radioactive contaminants. As a manager, you must work to prevent this. The key to protecting food is to make it as hard as possible for someone to tamper with it.

A food allergy is the body's negative reaction to a food protein. You should know about the most common food allergens. They include milk and dairy products; eggs and egg products; fish and shellfish; wheat; soy and soy products; and peanuts and tree nuts. Service staff must be able to tell customers about menu items that contain potential allergens. Kitchen staff must make sure that allergens are not transferred from food containing an allergen to the food served to a customer with allergies.

Chapter Review Case Study

Now take what you have learned in this chapter and apply it to the following case study.

Maria is a new assistant manager at a large family restaurant that has recently had some problems. Two weeks before Maria started, a customer who was allergic to dairy was served the house special, which contained milk. Luckily, the customer's reaction was not serious. However, since he had told the server he was allergic to dairy, he was mad about the incident. He even threatened to sue. The server, Kyle, was a new hire with a lot of potential. He felt so bad about the situation that he quit.

Also, several times in the last month, customers reported finding foreign objects in their food. This included a piece of plastic in some coleslaw, metal bits in the creamed spinach, and hair in the soup.

As Maria watched the staff during her first week, she saw several opportunities for improvement. First, she watched Lon, a buffet attendant. During the Sunday brunch buffet, Lon brought out a fresh pan of sausage links. Maria watched him use the spoon from the scrambled eggs to scoop the sausages from the old pan into the new. Then Lon sprayed some glass cleaner on the buffet's sneeze guard and wiped it down with a towel. After this, he took off his hat and retied his ponytail while standing by the steam table.

① What did Lon do wrong?

_____

_____

_____

_____

Later that same day, Maria overheard a conversation between Jean, a server, and a customer. The customer wanted to know what was in the barbeque sauce for the ribs, a specialty of the house. Jean cheerfully answered that the recipe was a secret. "I am allergic to peanuts," he replied, "so if it has any in it, I can't eat it." Jean said she didn't know, so the customer studied the menu some more. After a few moments, Jean told the customer she would come back for his order.

② What should Jean have done differently?

_____

_____

_____

_____

Several minutes later, Jean went back for the order. The customer asked for a plain hamburger—no bun or condiments—and fries. When he reminded her that he was allergic to peanuts, she wrote "allergic to peanuts" on the ticket and turned it in.

In the kitchen, Marc, the cook, went to work on the order. First, he checked the ingredients of the fries before starting the order. They were free of peanuts. Then he put the fries in a fryer that is only used for fries. Next, he grilled the burger. When everything was ready, he plated the food and put it in the pickup window.

③ What did Marc do wrong?

_____

_____

_____

_____

For answers, please turn to page 3.13.

Study Questions

Circle the best answer to each question below.

① **Eggs and peanuts are dangerous for people with which condition?**

A FAT TOM

B Food allergies

C Chemical sensitivity

D Poor personal hygiene

② **Cooking tomato sauce in a copper pot can cause which foodborne illness?**

A Hemorrhagic colitis

B Foodborne infection

C Toxic-metal poisoning

D Staphylococcal gastroenteritis

③ **To prevent chemical contamination, chemicals should be stored _____ food and utensils.**

A next to

B above

C separate from

D in the same area as

④ **Itching and tightening of the throat are symptoms of what?**

A Hepatitis A

B Food allergy

C Hemorrhagic colitis

D Ciguatera fish poisoning

⑤ **To prevent food allergens from being transferred to food,**

A clean and sanitize utensils before use.

B buy food from approved, reputable suppliers.

C store cold food at 41°F (5°C) or lower.

D avoid pewter tableware and copper cookware.

⑥ **What three points should a food defense program focus on to prevent possible threats to food?**

A Inspection reports, HACCP program, invoices

B Human element, building interior, building exterior

C Plant toxins, temperature logs, personal hygiene

D Cleaning schedules, labeling procedures, FAT TOM

For answers, please turn to page 3.14.

Answers

3.5   Can It Cause Contamination?

① Yes. The lost staple could have landed on food, equipment, or a prep area.

② No. He followed the manufacturer's directions in mixing the solution. He also sanitized the table while it was not being used.

③ Yes. Zinc from the tub could be transferred to the lemonade.

④ No. The food cannot touch the copper because of the stainless steel on the inside.

⑤ Yes. Glass could still be in the onions.

⑥ Yes. If the detergent container leaks, detergent would get into the stock pot.

⑦ Yes. Lead from the platter could be transferred to the food.

⑧ Yes. He did not label the smaller bottle. It could be mistaken for a different substance and contaminate food.

⑨ Yes. One of her false fingernails could come loose and get into the food. The false nails could also be a biological contaminant. False nails are hard to clean under and can hide dirt and other contaminants.

3.9   Identify the Symptoms

1, 5, 6, 7, 8, 11, and 14 should be marked.

3.9   Identify the Most Common Food Allergens

2, 3, 5, 6, 10, 14, 16, and 17 should be marked.

3.9   Keeping Food Safe for Customers with Allergies

① Wait to ask the chef. Suggest simpler items.

② Wash, rinse, and sanitize the utensils and equipment before making the order. Wash hands and change gloves before making the order.

3.10   Chapter Review Case Study

① Lon used the spoon from the eggs to handle the sausage. If a person with an allergy to eggs ate the sausage, he or she could have a reaction. Lon sprayed glass cleaner on the sneeze guard. This could contaminate the food on the buffet. He took off his hat and retied his ponytail near the food. This could contaminate the food with hair.

② Jean should have been aware of what was in the sauce. When she didn't know, she should have asked the cook or the manager to find out. She could have suggested a menu item that was made more simply.

Continued on the next page ▶

► *Continued from previous page*

③ Marc should have washed, rinsed, and sanitized the equipment used to prepare the order before starting on it. If the equipment was used to prepare food containing peanuts, it could transfer allergens. He should have washed his hands and changed his gloves before starting the order. If he touched food containing peanuts, his hands or gloves could transfer allergens.

3.12   **Study Questions**

① B

② C

③ C

④ B

⑤ A

⑥ B

Notes

4

The Safe Foodhandler

In the News

Desserts Outbreak

Fifteen members of a tour group got sick several hours after eating at a fine-dining restaurant. The tourists reported severe nausea, vomiting, and abdominal cramps. Health officials said that the symptoms and short onset pointed to the illness staphylococcal gastroenteritis.

The source of the illness was later identified as the Napoleon pastries that the group ate for dessert. Leftover pastries tested positive for *Staphylococcus aureus*. The restaurant's pastry chef also tested positive for the pathogen.

You Can Prevent This

The pastry chef in the story above spread the pathogen on his skin to the pastries he made. His customers got sick because he did not practice good personal hygiene. A good personal hygiene program is critical for preventing contamination of food by foodhandlers. The program should include policies that address the following areas.

- Avoiding personal behaviors that can contaminate food

- Washing and caring for hands

- Dressing for work and handling work clothes

- Limiting where employees can eat, drink, smoke, and chew gum or tobacco

- Preventing employees who may be carrying pathogens from working with or around food, or from working in the operation

---

Concepts from Earlier Chapters

Before reading this chapter, remember these concepts and facts.

Cross-contamination   Transfer of pathogens from one surface or food to another.

Virus   The leading cause of foodborne illness. Best controlled by practicing personal hygiene.

Bacteria   Pathogens that grow rapidly when conditions are right. Some produce toxins that make people sick.

Ready-to-eat food   Food that can be eaten without further prep, washing, or cooking.

How Foodhandlers Can Contaminate Food

At every step in the flow of food, foodhandlers can contaminate food. They might not even realize it when they do it. Something as simple as rubbing an ear while prepping a salad could make a customer sick. Even a foodhandler who appears to be healthy may spread foodborne pathogens. As a manager, you need to know the many ways foodhandlers can contaminate food.

Situations That Can Lead to Contaminating Food

Foodhandlers can contaminate food in any of the following situations.

- When they have a foodborne illness.

- When they have wounds that contain a pathogen.

- When they have contact with a person who is ill.

- When they touch anything that may contaminate their hands and then don't wash them. The buser in the photo at left may have contaminated his hands and could spread pathogens if he fails to wash them.

- When they have symptoms such as diarrhea, vomiting, or jaundice—a yellowing of the eyes or skin.

With some illnesses, a person may infect others before showing any symptoms. For example, a person could spread hepatitis A for weeks before having any symptoms. With other illnesses, a person may infect others for days or even months after symptoms are gone. Norovirus can be spread for days after symptoms have ended.

Some people carry pathogens and infect others without ever getting sick themselves. These people are called carriers. The bacteria *Staphylococcus aureus* is carried in the nose of 30 to 50 percent of healthy adults. About 20 to 35 percent of healthy adults carry it on their skin. If it is transferred from the nose or skin to food, which could happen in the photo at left, people can get sick.

Actions That Can Contaminate Food

People often do things that can spread pathogens without knowing it. To keep from causing a foodborne illness, pay close attention to what you do with your hands and avoid the following actions.

A Scratching the scalp

B Running fingers through the hair

C Wiping or touching the nose

D Rubbing an ear

E Touching a pimple or an infected wound

F Wearing a dirty uniform

G Coughing or sneezing into the hand

H Spitting in the operation

Apply Your Knowledge

Who Is at Risk?

Write an ✖ next to the foodhandler's name if there is a risk the foodhandler could spread pathogens.

① _____ Jamie, a prep cook, has a habit of rubbing his chin. Even though people tease him about this, he doesn't even notice when he touches it.

② _____ Rita, a pizza maker, has a bad headache but no fever. She gets a lot of headaches, but she always comes to work anyway.

③ _____ Lee, a sous chef, didn't have time to do laundry. He has to wear the same chef coat he wore yesterday.

④ _____ Phillip, a grill operator, has a small cut on his cheek. It's not bleeding, but he has a bandage on it.

⑤ _____ Gary, a dishwasher, has allergies. Sometimes he needs to spit, so he spits in the garbage can next to the sink.

⑥ _____ Helen's children have had diarrhea. Her mother has been caring for them so Helen, a line cook, can go to work.

⑦ _____ Victor, an ice cream server, likes outdoor activities. Last weekend, he went camping in an area where there were no indoor toilets.

⑧ _____ Sabrina, a pastry chef, has dandruff, which itches. She tries not to scratch her head, but sometimes she just has to do it.

For answers, please turn to page 4.19.

A Good Personal Hygiene Program

To keep foodhandlers from contaminating food, your operation needs a good personal hygiene program. As a manager, you must make sure this program succeeds. You must create and support policies that address the following areas.

A Hand practices

- Handwashing
- Hand care
- Glove use
- Bare-hand contact with ready-to-eat food

B Personal cleanliness

C Clothing—including hair restraints and jewelry

Employees must also avoid certain habits and actions, maintain good health, cover wounds, and report illnesses.

Managing a Personal Hygiene Program

Don't underestimate your role in a personal hygiene program. You have many responsibilities to help make the program work.

- Creating personal hygiene policies.
- Training foodhandlers on those policies and retraining them when necessary.
- Modeling the right behavior at all times. The manager in the photo at left is modeling good personal hygiene practices. He is wearing clean clothes and a hair restraint. He is also using gloves.
- Supervising food safety practices at all times.
- Revising personal hygiene policies when laws or science change.

Handwashing

Handwashing is the most important part of personal hygiene. It may seem like an obvious thing to do. Even so, many foodhandlers do not wash their hands the right way or as often as they should. You must train your foodhandlers to wash their hands and then you must monitor them.

How to Wash Hands

To wash hands the right way, follow the steps shown below. The whole process should take about 20 seconds.

❶ Wet hands and arms.

Use running water as hot as you can comfortably stand. It should be at least 100°F (38°C).

❷ Apply soap.

Apply enough to build up a good lather.

❸ Scrub hands and arms vigorously.

Scrub them for 10 to 15 seconds. Clean under fingernails and between fingers.

❹ Rinse hands and arms thoroughly.

Use running water.

❺ Dry hands and arms.

Use a single-use paper towel or a hand dryer.

If you are not careful, you can contaminate your hands after washing them. Consider using a paper towel to turn off the faucet and to open the door when leaving the restroom.

PATHOGEN PREVENTION

When to Wash Hands

Foodhandlers must wash their hands before they start work. They must also do it **after** the following activities.

- Using the restroom. Foodhandlers with foodborne illnesses such as Norovirus gastroenteritis can transfer the pathogen to food if they don't wash their hands after using the restroom.

- Handling raw meat, poultry, and seafood (before *and* after).

- Touching the hair, face, or body.

- Sneezing, coughing, or using a tissue.

- Eating, drinking, smoking, or chewing gum or tobacco.

- Handling chemicals that might affect food safety.

- Taking out garbage.

- Clearing tables or busing dirty dishes.

- Touching clothing or aprons.

- Handling money.

- Touching anything else that may contaminate hands, such as dirty equipment, work surfaces, or towels. The foodhandler in the photo at left should wash his hands after using the towel to wipe the prep counter.

Hand Antiseptics

Hand antiseptics are liquids or gels that are used to lower the number of pathogens on skin. If used, they must comply with Food and Drug Administration (FDA) standards.

Only use hand antiseptics **after** handwashing. NEVER use them in place of it. Wait for a hand antiseptic to dry before you touch food or equipment.

Check your local regulatory requirements.

How This Relates to Me

Are hand antiseptics allowed by your local regulatory authority?

_____ Yes   _____ No

If allowed, what are the regulatory requirements?

_____

_____

_____

Hand Care

In addition to washing, hands need other care to prevent spreading pathogens. Make sure foodhandlers follow these guidelines.

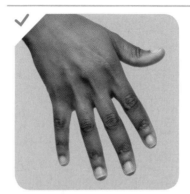

✓ **Fingernail length**   Keep fingernails short and clean. Long fingernails may be hard to keep clean.

✗ **False fingernails**   Do **NOT** wear false fingernails. They can be hard to keep clean. False fingernails also can break off into food. Some local regulatory authorities allow false nails if single-use gloves are worn.

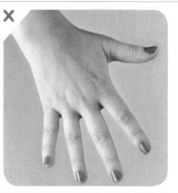

✗ **Nail polish**   Do **NOT** wear nail polish. It can disguise dirt under nails and may flake off into food. Some regulatory authorities allow polished nails if single-use gloves are worn.

PATHOGEN PREVENTION

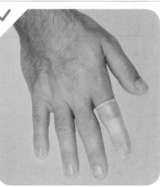

✓ **Hand wounds**   Wear a bandage over wounds on hands and arms. Make sure it keeps the wound from leaking.

You must wear a single-use glove or finger cot (a finger cover) over bandages on hands or fingers. These will protect the bandage and keep it from falling off into food. It will also keep wounds that contain pathogens like *Staphylococcus aureus* from contaminating food and causing illnesses.

Check your local regulatory requirements.

Something to Think About...

More Than They Bargained For

At a restaurant on the East Coast, the salad bar was very popular. One afternoon while prepping the lettuce, an employee cut her finger. She bandaged it right away and returned to work. While she was tossing the salad, the bandage fell off into the lettuce.

A short time later, a customer reported that she had found a used bandage in her salad. The manager apologized and quickly comped her meal. Fortunately, the customer was easygoing, and the rest of the evening was uneventful.

Single-Use Gloves

Single-use gloves can help keep food safe by creating a barrier between hands and food. But NEVER use gloves in place of handwashing. Hands must be washed before putting on gloves and when changing to a new pair.

Buying Gloves

When buying gloves for handling food, follow these guidelines.

Disposable gloves   Buy only single-use gloves for handling food. NEVER wash and reuse gloves.

Multiple sizes   Make sure you provide different glove sizes. Gloves that are too big will not stay on. Those that are too small will tear or rip easily. The photo at left shows a correct fit.

Latex alternatives   Some foodhandlers and customers may be sensitive to latex. Consider providing gloves made from other materials.

When to Change Gloves

Foodhandlers must change gloves at all of these times.

* As soon as they become soiled or torn

* Before beginning a different task

* At least every four hours during continual use, and more often if necessary

* After handling raw meat, seafood, or poultry and before handling ready-to-eat food

Check your local regulatory requirements.

How This Relates to Me

What are your local regulatory requirements for glove use?

_____

_____

_____

Bare-Hand Contact with Ready-To-Eat Food

Your foodhandlers may be allowed to handle ready-to-eat food with bare hands, as the prep chef in the photo at left is doing. But it may increase the risk of contaminating the food. If your area allows bare-hand contact with ready-to-eat food, you must have policies for employee health and train employees in handwashing and personal hygiene practices.

Check your local regulatory requirements.

How This Relates to Me

Does your local regulatory authority allow bare-hand contact with ready-to-eat food?

_____ Yes   _____ No

If allowed, what are the regulatory requirements?

_____

_____

_____

Personal Cleanliness

Pathogens can be found on hair and skin that are not kept clean. These pathogens can be transferred to food and food equipment. Make sure foodhandlers shower or bathe before work.

Work Attire

Foodhandlers in dirty clothes may give a bad impression of your operation. More important, dirty clothing may carry pathogens that can cause foodborne illnesses. Set up a dress code and make sure all employees follow it. The code should include the following guidelines.

Hair restraints   Wear a clean hat or other hair restraint. Foodhandlers with facial hair should also wear a beard restraint.

Clean clothing   Wear clean clothing daily. If possible, change into work clothes at work. Dirty clothing that is stored in the operation must be kept away from food and prep areas. This includes dirty aprons, chef coats, and uniforms.

Aprons   Remove aprons when leaving prep areas. For example, aprons should be removed and stored properly before taking out garbage or using the restroom.

Jewelry   Remove jewelry from hands and arms before prepping food or when working around prep areas. You cannot wear any of the following.

- Rings, except for a plain band
- Bracelets, including medical bracelets
- Watches

Your company may also require you to remove other types of jewelry. This may include earrings, necklaces, and facial jewelry. Servers may wear jewelry if allowed by company policy.

Check your local regulatory requirements.

How This Relates to Me

What are your local regulatory requirements for hair restraints, clothing, jewelry, and aprons?

_____

_____

Policies for Eating, Drinking, Smoking, and Chewing Gum or Tobacco

Small droplets of saliva can contain thousands of pathogens. In the process of eating, drinking, smoking, or chewing gum or tobacco, saliva can be transferred to hands or directly to food being handled.

Do NOT eat, drink, smoke, or chew gum or tobacco at any of these times.

- When prepping or serving food
- When working in prep areas
- When working in areas used to clean utensils and equipment

Only eat, drink, smoke, and chew gum or tobacco in designated areas.

Some regulatory authorities allow foodhandlers to drink from a covered container with a straw while in prep and dishwashing areas.

Check your local regulatory requirements.

How This Relates to Me

Does your local regulatory authority allow employees to drink in prep and dishwashing areas?

_____ Yes   _____ No

If so, what are the regulatory requirements?

_____

_____

Policies for Reporting Health Issues

You must encourage your foodhandlers to report any health problems before they come to work. They should also let you know right away if they get sick while working, as the employee in the photo at left is doing.

When foodhandlers are ill, you may need to keep—or *restrict*—them from working with or around food. Sometimes, you may need to keep—or *exclude*—them from coming in to work. Use the following information to help you decide how to handle foodhandler illnesses.

Handling Foodhandler Illnesses

If	Then
The foodhandler has a sore throat with a fever.	Restrict the foodhandler from working with or around food.
	Exclude the foodhandler from the operation if you primarily serve a high-risk population.
The foodhandler has at least one of these symptoms. • Vomiting • Diarrhea • Jaundice	Exclude the foodhandler from the operation. Before returning to work, foodhandlers who vomited or had diarrhea must meet one of these requirements. • Have had no symptoms for at least 24 hours • Have a written release from a medical practitioner Foodhandlers with jaundice must have a written release from a medical practitioner before they can go back to work.
The foodhandler has been diagnosed with a foodborne illness caused by one of these pathogens. • *Salmonella* Typhi • *Shigella* spp. • Shiga toxin-producing *E. coli* • Hepatitis A • Norovirus	Exclude the foodhandler from the operation. Notify the local regulatory authority. Work with the foodhandler's medical practitioner and/or the local regulatory authority to decide when the person can go back to work.

Check your local regulatory requirements.

How This Relates to Me

When does your regulatory authority require foodhandlers to be restricted from working with or around food?

_____

_____

What other illnesses would require you to exclude a foodhandler from your operation?

_____

_____

Something to Think About...

A Shaky Situation

Sixteen people became sick and four were hospitalized after drinking milk shakes contaminated with shiga toxin-producing _E. coli_ at a drive-in restaurant. The outbreak was started by an employee who came to work even though she was ill. The employee made a milk-shake mix that was served over a five-day period. She had diarrhea and severe abdominal cramps before making the shake mix. However, she continued to work and was later found to be ill with shiga toxin-producing _E. coli_.

Three of the sick customers were kept in the hospital for short periods. A fourth, a 15-year-old girl, required dialysis. The drive-in was temporarily closed but was cleaned and reopened a few days later.

Apply Your Knowledge

Check Your Handwashing Savvy

Circle the letters of the right steps for handwashing from column A. Then put them in the right order in column B.

Column A

Ⓐ Scrub hands and arms for 3 to 5 seconds.

Ⓑ Scrub hands and arms for 10 to 15 seconds.

Ⓒ Rinse hands and arms in running water.

Ⓓ Rinse hands and arms in warm, standing water.

Ⓔ Wet hands and arms with water at least 100°F (38°C).

Ⓕ Wet hands and arms with water at least 115°F (46°C).

Ⓖ Apply enough soap to build up a good lather.

Ⓗ Apply enough soap to cover the palm of your hand.

Ⓘ Dry hands and arms on a shared towel.

Ⓙ Dry hands and arms with a single-use paper towel or hand dryer.

Column B

① _____

② _____

③ _____

④ _____

⑤ _____

For answers, please turn to page 4.19.

Apply Your Knowledge

When to Wash Hands?

Paul was just promoted from buser to prep-cook-in-training at the busy family restaurant where he has worked for nine months. Since his promotion last week, he has already learned some basic tasks. Today, Paul arrived promptly at work at 8:00 a.m. and punched in. In a preshift meeting, his manager, Miguel, told him his first task was to debone 20 pounds of raw chicken breasts. After that, Paul was to work with Linda, an experienced prep cook.

Eager to make a good impression, Paul got the raw chicken from the cooler. He put on single-use gloves and started removing the bones from the breasts. About halfway through, Miguel stopped by with someone in a suit, whom Miguel introduced as one of the new owners. Paul took off his gloves to shake the owner's hand. Afterwards, he put on a new pair of single-use gloves and got back to the chicken.

After Paul finished the chicken, he put it in the cooler. Then he noticed a pile of dirty dishes next to the dishwasher. He decided to help out and loaded the dishes into a dish rack. When it was full, he ran the load.

Linda came by and told him to prep vegetables for the salad bar. The salad bar needed to be set up in an hour, so Paul put on single-use gloves and got started right away. As he was chopping lettuce, Paul suddenly sneezed. Fortunately, he was able to turn away from the prep table in time. He also was able to catch the sneeze with his hand. He changed gloves and went back to chopping the lettuce. Then Paul needed a tissue. Not having one handy, he had to go to the bathroom to blow his nose on toilet paper. While he was there, he also used the toilet. Then he returned to the prep area, put on new gloves, and finished prepping vegetables. He placed the prepped vegetables in the cooler.

Linda asked Paul to let the salad bar attendant know the vegetables were ready. Then she asked him to work with her on the grill during the lunch hour. She started him off by showing him how to make a grilled chicken sandwich. When a second order for a grilled chicken sandwich came in, Paul got to make it. A bit nervous, he dropped the tongs for the raw chicken before he could get a piece onto the grill. Linda went to get a clean pair of tongs but got delayed by a conversation with Miguel.

Paul was sure he could make the sandwich without Linda's help. While she was talking with Miguel, he picked up a raw chicken breast with his hands and put it on the grill. Then he got out a fresh bun and put it in the toaster. Before he could finish the sandwich, Linda came back, and they finished it together. As they worked together on the orders, the lunch rush flew by.

When should Paul have washed his hands?

_____

_____

_____

_____

Exclusion or Restriction?

Write an E next to the statement if the foodhandler must be excluded from the operation. Write an R next to the statement if the person should be restricted from working with or around food.

① _____ Bill, a line cook at a family restaurant, has a sore throat with a fever.

② _____ Joe, a prep cook, has diarrhea.

③ _____ Mary, a sous chef, has been diagnosed with hepatitis A.

For answers, please turn to page 4.19.

Chapter Summary

Foodhandlers can spread pathogens and contaminate food at every step in the flow of food. So good personal hygiene is critical in an operation. Hands must be washed the right way. This is especially important before starting work; after using the restroom; after sneezing, coughing, smoking, eating, or drinking; and before and after handling raw meat, poultry, and fish. Hands also need other care. Fingernails should be kept short and clean. Wounds on hands and arms should be covered with clean bandages. Hand cuts should be covered with gloves or finger cots.

Before handling food or working in prep areas, foodhandlers must put on clean clothing and a clean hair restraint. They must remove jewelry from hands and arms. Aprons should be removed and stored when foodhandlers leave prep areas. They also should report health problems to you before working with food. Foodhandlers must not work if they have been diagnosed with a foodborne illness caused by *Salmonella* Typhi; *Shigella* spp.; shiga toxin-producing *E. coli*; hepatitis A; or Norovirus. Foodhandlers also must not come to work if they have symptoms that include diarrhea, vomiting, or jaundice. Staff should not work with or around food if they have a sore throat and a fever.

To keep foodhandlers from contaminating food, your operation needs a good personal hygiene program. You can minimize the risk of foodborne illnesses by establishing a program, training, and enforcing it. Most important, you must set an example yourself by practicing good personal hygiene.

Chapter Review Case Study

Now take what you have learned in this chapter and apply it to the following case study.

Randall is a foodhandler at a deli. It is 7:47 a.m., and he has just woken up. He is scheduled to be at work and ready to go by 8:00 a.m. When he gets out of bed, his stomach feels queasy. He blames that on the beer he had the night before. Fortunately, Randall lives only five minutes from work. Despite this, he doesn't have enough time to take a shower. He grabs the same uniform he wore the day before when prepping chicken. He also puts on his watch and several rings.

Randall does not have luck on his side today. On the way to the restaurant, his oil light comes on. He is forced to pull off the road and add oil to his car. When he gets to work, he realizes he has left his hat at home. Randall is greeted by an angry manager. The manager puts Randall to work right away, loading the rotisserie with raw chicken. Randall then moves to serving a customer who orders a freshly made salad. Randall is known for his salads and makes the salad to the customer's approval.

The manager asks Randall to take out the garbage and then make potato salad for the lunch-hour rush. On the way back from the garbage run, Randall tells the manager that his stomach is bothering him. The manager, thinking of his staff shortage, asks Randall to stick it out as long as he can. Randall agrees and gets out the ingredients for the potato salad. Then he heads to the restroom in hope of relieving his symptoms. After quickly rinsing his hands in the restroom, he finds that the paper towels have run out. Short of time, he wipes his hands on his apron.

Later, Randall cuts his finger while making the potato salad. He bandages the cut and continues his prep work. The manager then tells Randall to clean the few tables in the deli that are available for customers. He puts on a pair of single-use gloves and cleans and sanitizes the tables. When finished, Randall grabs a piece of chicken from the rotisserie for a snack. He takes the chicken with him to the prep area, so he can get back to making the potato salad.

Randall and his manager made many errors. Identify as many as you can on a separate sheet of paper.

- If you can identify only 8 to 12 errors, you may need to reread this chapter.
- If you can identify 13 to 16 errors, you have a good understanding of this chapter.
- If you can identify more than 16 errors, you are on your way to becoming a food safety expert.

For answers, please turn to page 4.20.

Study Questions

Circle the best answer to each question below.

① What must foodhandlers do after touching their hair, face, or body?

A Wash their hands

B Rinse their gloves

C Change their aprons

D Use a hand antiseptic

② What should foodhandlers do after prepping food and before using the restroom?

A Wash their hands

B Take off their hats

C Change their gloves

D Take off their aprons

③ Which piece of jewelry can be worn by a foodhandler?

A Diamond ring

B Medical bracelet

C Plain band ring

D Watch

④ When should hand antiseptics be used?

A Before washing hands

B After washing hands

C In place of washing hands

D In place of wearing gloves

⑤ When should foodhandlers who wear gloves wash their hands?

A After putting on the gloves

B Before taking off the gloves

C After applying a hand antiseptic

D Before putting on the gloves

⑥ Foodhandlers should keep their fingernails

A short and unpolished.

B long and unpolished.

C long and painted with nail polish.

D short and painted with nail polish.

⑦ **A cook wore single-use gloves while forming raw ground beef into patties. The cook continued to wear them while slicing hamburger buns. What mistake was made?**

A The cook did not wear reusable gloves while handling the raw ground beef and hamburger buns.

B The cook did not clean and sanitize the gloves before handling the hamburger buns.

C The cook did not wash hands before putting on the same gloves to slice the hamburger buns.

D The cook did not wash hands and put on new gloves before slicing the hamburger buns.

⑧ **When a foodhandler has been diagnosed with shigellosis, what steps must be taken?**

A The foodhandler must be told to not come in to work.

B The foodhandler must be given a nonfoodhandling position.

C The foodhandler can work, but must wear gloves when handling food.

D The foodhandler can work, but must wash hands every 15 minutes.

⑨ **Foodhandlers can't work in their operation if they have an illness caused by which pathogen?**

A *Vibrio vulnificus*

B *Salmonella* Typhi

C *Clostridium botulinum*

D *Clostridium perfringens*

⑩ **Foodhandlers who work in a nursing home can't work in the operation if they have which symptom?**

A Thirst with itching

B Sore throat with fever

C Soreness with fatigue

D Headache with soreness

⑪ **Foodhandlers should not eat, drink, smoke, or chew gum or tobacco while**

A bare handed.

B on their break.

C prepping food.

D counting money.

Continued on the next page ▶

▶ *Continued from previous page*

⑫ **What should foodhandlers do if they cut their fingers while prepping food?**

A Cover the wound with a bandage.

B Stay away from food and prep areas.

C Cover the hand with a glove or a finger cot.

D Cover the wound with a bandage and a glove or a finger cot.

⑬ **What should a manager at a nursing home do if a cook calls in with a headache, nausea, and diarrhea?**

A Tell the cook to stay away from work and see a doctor.

B Tell the cook to rest for a couple of hours and then come to work.

C Tell the cook to come in for a couple of hours and then go home.

D Tell the cook to go to the doctor and then immediately come to work.

For answers, please turn to page 4.20.

Answers

4.3 **Who Is at Risk?**

1, 3, 5, 6, and 8 should be marked.

4.13 **Check Your Handwashing Savvy**

① E

② G

③ B

④ C

⑤ J

4.14 **When to Wash Hands?**

Paul should have washed his hands at the following times.

- Before getting the chicken from the cooler
- Before putting on the first pair of gloves to debone the chicken
- After shaking hands with the owner and before putting on the second pair of gloves
- After finishing deboning chicken
- After loading the dishes into the dishwasher
- Before putting on gloves and handling the vegetables for the salad bar
- After sneezing and before putting on the third pair of gloves
- After using the restroom
- Before putting on the fourth pair of gloves and continuing to prep vegetables
- Before working on the grill
- Before touching the raw chicken breast
- After placing the raw chicken breast on the grill and before handling the fresh bun

4.14 **Exclusion or Restriction?**

① R

② E

③ E

Continued on the next page ▶

► *Continued from previous page*

4.15   Chapter Review Case Study

Randall made 17 errors.

- Randall did not take a bath or shower before work.

- Randall wore a dirty uniform to work.

- Randall should have removed his watch and rings (with the exception of a plain band) before prepping and serving food.

- Randall did not wear a hair restraint.

- Randall did not report his illness to the manager before coming to work.

- Randall did not wash his hands before handling the raw chicken.

- Randall did not wash his hands after handling the raw chicken.

- The manager did not ask about Randall's symptoms. If Randall were to report that he had diarrhea, the manager should have sent him home.

- Randall did not wash his hands correctly after taking out the garbage.

- Randall did not wash his hands correctly after using the restroom.

- Randall did not dry his hands correctly after washing them. He got them dirty again when he wiped them on his apron.

- Randall wore his apron into the restroom.

- The manager did not make sure the restroom was stocked with paper towels.

- Randall did not wear a finger cot or a single-use glove over the bandaged finger.

- Randall did not wash his hands before putting on the single-use gloves.

- Randall touched the ready-to-eat chicken with his contaminated gloves.

- Randall was eating chicken while prepping food.

4.16   Study Questions

① A   ⑧ A

② D   ⑨ B

③ C   ⑩ B

④ B   ⑪ C

⑤ D   ⑫ D

⑥ A   ⑬ A

⑦ D

Notes

II The Flow of Food Through the Operation

5

The Flow of Food: An Introduction

In the News

Deli Cited for Temperature Violations

A major city's health department cited a local deli for not checking food temperatures. According to the inspector, "There was sausage being held at 88°F (31°C), hash browns at 52°F (11°C), rice pudding at 48°F (8°C), and cheesecake at 52°F (11°C)."

The health inspector worked with the manager of the operation to help identify steps for taking temperatures. One step included establishing regular monitoring times and keeping a temperature log.

The deli manager confirmed that the operation was correcting the problem. He added that the health department had scheduled another inspection for the following week.

You Can Prevent This

As you can see in the story above, controlling time and temperature is critical for keeping food safe. But you must also prevent pathogens from spreading throughout the operation. In this chapter, you will learn about the following tools and practices to help you keep food safe.

- Preventing cross-contamination
- Preventing time-temperature abuse
- Using the right kinds of thermometers to take temperatures
- Keeping your thermometers accurate

Concepts from Earlier Chapters

Before reading this chapter, remember these concepts and facts.

Cross-contamination   Transfer of pathogens from one surface or food to another.

Time-temperature abuse   When food stays too long at temperatures that are good for pathogen growth.

Temperature danger zone   Temperature range between 41°F and 135°F (5°C to 57°C). Foodborne pathogens grow well in it.

TCS food   Food that requires time and temperature control for safety.

Hazards in the Flow of Food

To keep food safe, you must apply what you learn in the ServSafe program throughout the flow of food. This requires a good understanding of how to prevent cross-contamination and time-temperature abuse.

The Flow of Food

The path that food takes through your operation is called the flow of food. It begins when you buy the food and ends when you serve it.

$	Purchasing	chapter 6
🚚	Receiving	chapter 6
	Storing	chapter 6
🔪	Preparation	chapter 7
	Cooking	chapter 7
🕐	Holding	chapter 8
❄	Cooling	chapter 7
	Reheating	chapter 7
🍽	Serving	chapter 8

You are responsible for the safety of the food at every point in this flow—and many things can happen to it.

For example, a frozen food might be safe when it leaves the processor's plant. However, on the way to the supplier's warehouse, the food might thaw. When you receive it, you might not notice there's anything wrong with it. Once in your operation, the food might not be stored correctly, or it might not be cooked to the right internal temperature. These mistakes can add up and cause a foodborne illness.

Cross-Contamination

Pathogens can move around easily in your operation. They can be spread from food or unwashed hands to prep areas, equipment, utensils, or other food.

Cross-contamination can happen at almost any point in the flow of food. When you know how and where it can happen, it is fairly easy to prevent. The most basic way is to keep raw and ready-to-eat food away from each other. Here are some guidelines for doing this.

Separating equipment   Use separate equipment for each type of food. For example, use one set of cutting boards, utensils, and containers for raw poultry. Use another set for raw meat. Use a third set for produce. Colored cutting boards and utensil handles can help keep equipment separate. The color tells foodhandlers which equipment to use with each food item. You might use yellow for raw chicken, red for raw meat, and green for produce, as the prep chef is doing in the photo at left.

PATHOGEN PREVENTION

Cleaning and sanitizing   Clean and sanitize all work surfaces, equipment, and utensils after each task. When you cut up raw chicken, for example, you cannot get by with just rinsing the equipment. To prevent pathogens such as *Salmonella* spp. from contaminating food, you must wash, rinse, and sanitize the equipment. Know which cleaners and sanitizers to use for each job. See chapter 11 for more information on cleaning and sanitizing.

Prepping food at different times   If you need to use the same table to prep different types of food, prep raw meat, fish, and poultry and ready-to-eat food at different times. You must clean and sanitize work surfaces and utensils between each product. For example, by prepping ready-to-eat food before raw food, you can minimize the chance for cross-contamination.

Buying prepared food   Buy food items that don't require much prepping or handling. For example, you could buy precooked chicken breasts or chopped lettuce, as shown in the photo at left.

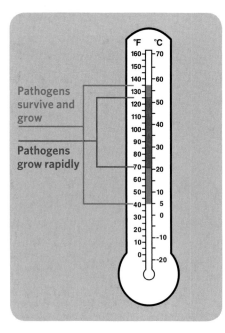

Time-Temperature Abuse

Most foodborne illnesses happen because TCS food has been time-temperature abused. Remember, food has been time-temperature abused any time it remains at 41°F to 135°F (5°C to 57°C). This is called the temperature danger zone, because pathogens grow in this range. But they grow much faster at 70°F to 125°F (21°C and 52°C). These ranges are shown at left.

Food is being temperature abused whenever it is handled in the following ways.

- Cooked to the wrong internal temperature
- Held at the wrong temperature
- Cooled or reheated incorrectly

The longer food stays in the temperature danger zone, the more time pathogens have to grow. To keep food safe, you must reduce the time it spends in this temperature range. If food is held in this range for four or more hours, you must throw it out.

Avoiding Time-Temperature Abuse

Employees should avoid time-temperature abuse by following good policies and procedures. They should cover the following areas.

Monitoring   Learn which food items should be checked, how often, and by whom. Then assign duties to foodhandlers in each area. Make sure they understand what to do, how to do it, and why it is important. For example, the manager in the photo at left is making sure the cook can check the temperature of a chicken breast.

Tools   Make sure the right kinds of thermometers are available. Give foodhandlers their own thermometers. Have them use timers in prep areas to check how long food is in the temperature danger zone.

Recording   Have foodhandlers record temperatures regularly, as the chef is doing in the photo at left. Make sure they write down when the temperatures were taken. Print simple forms for recording this information. Post them on clipboards outside of coolers and freezers, near prep areas, and next to cooking and holding equipment.

Time and temperature control   Have procedures to limit the time food spends in the temperature danger zone. This might include limiting the amount of food that can be removed from a cooler when prepping it.

Corrections   Make sure foodhandlers know what to do when time and temperature standards are not met. For example, if you hold soup on a steam table and its temperature falls below 135°F (57°C) after two hours, you might reheat it to the correct temperature.

Apply Your Knowledge

An Ounce of Prevention...

Place a ✔ next to the practice if it helps prevent cross-contamination.

① _____   Use separate cutting boards for prepping raw meat and raw vegetables.

② _____   Wash and rinse a cutting board after prepping raw fish.

③ _____   Buy diced onions instead of dicing them in the operation.

④ _____   Prep raw chicken and potato salad on the table at the same time.

Is It Safe?

Read each story and decide if the foodhandler handled the food safely. Explain why or why not in the space provided.

Anita had to prepare six tuna salad sandwiches. She went to the cooler and pulled out a large hotel pan of tuna salad and put it on the prep table. She was interrupted several times to help with other tasks. After assembling the sandwiches, she covered the pan of tuna salad, dated it, and put it back in the cooler.

① Did Anita handle the food safely? Why or why not?

_____

_____

_____

_____

_____

_____

_____

_____

_____

Jerry cut up raw chickens on a cutting board on the prep table. Then he washed and rinsed the table and equipment he used. After that, he sliced onions and peppers on the same cutting board on the prep table. Before he left for the day, he washed, rinsed, and sanitized the prep table and equipment.

② Did Jerry handle the food safely? Why or why not?

_____

_____

_____

_____

_____

_____

_____

_____

_____

For answers, please turn to page 5.16.

Monitoring Time and Temperature

To keep food safe, you must control the amount of time it spends in the temperature danger zone. This requires monitoring. The most important tool you have to monitor temperature is the thermometer. Three types are commonly used in operations.

- Bimetallic stemmed thermometers
- Thermocouples
- Thermistors

The infrared thermometer, while not as common, is becoming more popular. Other tools are available that can help you monitor both time and temperature. A time-temperature indicator is one type.

All of these tools will be effective only if you follow specific guidelines for using them. Tools also have to be adjusted regularly to keep them accurate.

Bimetallic Stemmed Thermometer

A bimetallic stemmed thermometer, shown in the photo at left, can check temperatures from 0°F to 220°F (–18°C to 104°C). This makes it useful for checking temperatures during the flow of food. For example, you can use it to check food temperatures during receiving. You can also use it to check food in a hot- or cold-holding unit.

A bimetallic stemmed thermometer measures temperature through its metal stem. When checking temperatures, insert the stem into the food up to the dimple. You must do this because the sensing area of the thermometer goes from the tip of the stem to the dimple. This trait makes this thermometer useful for checking the temperature of large or thick food. It is usually not practical for thin food such as hamburger patties.

If you buy these thermometers for your operation, make sure they have these features.

Calibration nut You can adjust the thermometer to make it accurate by using its calibration nut.

Easy-to-read markings Clear markings reduce the chance that someone will misread the thermometer.

Dimple The dimple is the mark on the stem that shows the end of the temperature-sensing area.

Accuracy Make sure the manufacturer guarantees accuracy to within ±2°F (±1°C).

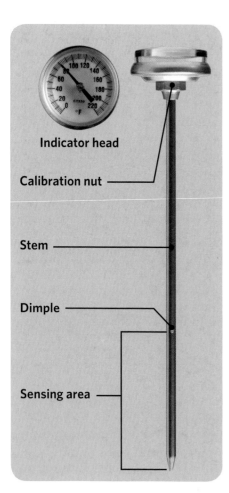

Indicator head

Calibration nut

Stem

Dimple

Sensing area

Thermocouples and Thermistors

Thermocouples, such as the one in the photo at left, and thermistors are also common in operations. They measure temperatures through a metal probe. Temperatures are displayed digitally. The sensing area on thermocouples and thermistors is on the tip of their probe. This means you don't have to insert them into the food as far as bimetallic stemmed thermometers to get a correct reading. Thermocouples and thermistors are good for checking the temperature of both thick and thin food.

Thermocouples and thermistors come in several styles and sizes. Many come with different types of probes. The photos below show some basic types.

Immersion probes Use these to check the temperature of liquids. This could include soups, sauces, and frying oil.

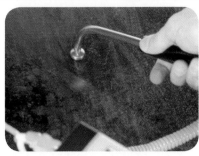

Surface probes Use these to check the temperature of flat cooking equipment, such as griddles.

Penetration probes Use these to check the internal temperature of food. They are especially useful for checking the temperatures of thin food, such as hamburger patties or fish fillets.

Air probes Use these to check the temperature inside coolers and ovens.

Infrared (Laser) Thermometers

Infrared thermometers measure the temperatures of food and equipment surfaces. For example, the foodhandler in the photo at left is using one to measure the temperature of a grill top. These thermometers are quick and easy to use.

Infrared thermometers do not need to touch a surface to check its temperature. This means there is less chance for cross-contamination and damage to food. However, these thermometers cannot measure air temperature or the internal temperature of food.

Follow these guidelines for using infrared thermometers.

Distance   Hold the thermometer as close to the food or equipment as you can without touching it.

Barriers   Remove anything between the thermometer and the food, food package, or equipment. Do NOT take readings through metal, such as stainless steel or aluminum. Do NOT take readings through glass.

Manufacturer's directions   Always follow the manufacturer's guidelines. This should give you the most accurate readings.

Other Temperature-Recording Devices

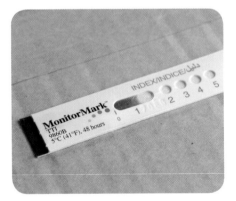

Some devices monitor both time and temperature. The time-temperature indicator (TTI), shown in the photo at left, is an example. These tags are attached to packaging by the supplier. A color change appears in the window if the food has been time-temperature abused during shipment or storage. This color change is not reversible, so you know if the food has been abused.

Some suppliers place temperature-recording devices inside their delivery trucks. These devices constantly check and record temperatures. You can check the device during receiving to make sure food was at safe temperatures while it was being shipped.

How to Calibrate Thermometers

Thermometers can lose their accuracy when they are bumped or dropped. It can also happen when they go through severe temperature change. When this happens, the thermometer must be calibrated, or adjusted, to give a correct reading. Most thermometers can be calibrated, but some can't and must be replaced or sent back to the manufacturer. You should follow the manufacturer's directions.

There are two ways to calibrate a thermometer. One is to adjust it based on the temperature at which water freezes. This is called the ice-point method. The other way is to adjust it based on the temperature at which water boils. This is called the boiling-point method. While either way works, the ice-point method is more common.

Ice-Point Method for Calibrating a Thermometer

Follow these steps to calibrate a thermometer using the ice-point method. The photos below show a bimetallic stemmed thermometer being calibrated.

❶ Fill a large container with crushed ice. Add tap water until the container is full.

Note: Stir the mixture well.

❷ Put the thermometer stem or probe into the ice water. Make sure the sensing area is under water.

Wait 30 seconds, or until the indicator stops moving. On thermocouples and thermistors, wait until the readout stops moving.

Note: Do NOT let the probe touch the container.

❸ Adjust the thermometer so it reads 32°F (0°C).

How you do this depends on the type of thermometer being used.

Bimetallic stemmed thermometers   Hold the calibration nut with a wrench or other tool. Rotate the thermometer head until it reads 32°F (0°C).

Thermocouples and thermistors   Follow the manufacturer's directions. On some devices, you can press a reset button.

Boiling-Point Method for Calibrating a Thermometer

Follow these steps to calibrate a thermometer using the boiling-point method. The photos below show a bimetallic stemmed thermometer being calibrated.

1 Bring tap water to a boil in a deep pan.

2 Put the thermometer stem or probe into the boiling water. Make sure the sensing area is under water.

Wait 30 seconds, or until the indicator stops moving. On thermocouples and thermistors, wait until the readout stops moving.

Note: Do NOT let the probe touch the container.

3 Adjust the thermometer so it reads 212°F (100°C).

Note: This temperature will vary depending on the boiling point for your elevation.

How you do this depends on the type of thermometer being used.

Bimetallic stemmed thermometers Hold the calibration nut with a wrench or other tool. Rotate the thermometer head until it reads 212°F (100°C).

Thermocouples and thermistors Follow the manufacturer's directions. On some devices, you can press a reset button.

General Thermometer Guidelines

You should know how to use and care for each type of thermometer in your operation. In general, follow the guidelines below. However, you should always follow manufacturers' directions.

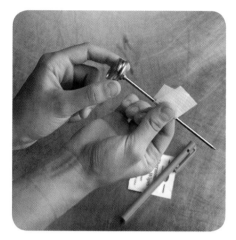

Cleaning and sanitizing   Thermometers must be washed, rinsed, sanitized, as seen in the photo at left, and air-dried. Keep storage cases clean too. Do these things before and after using thermometers to prevent cross-contamination. Be sure the sanitizing solution you use is for food-contact surfaces. Always have plenty of clean and sanitized thermometers on hand.

Calibration   Make sure your thermometers are accurate by calibrating them regularly. Aside from the times mentioned earlier, you should do this before each shift. You should also do this before the first delivery arrives.

Accuracy   Some types of thermometers cannot be calibrated. This includes most hanging thermometers. Those kept in coolers and freezers are often bumped or dropped, which can cause them to lose accuracy. Check the air temperature in these units using a thermocouple with an air probe. Then compare the reading to that of the hanging thermometer. Replace it if the readings don't match.

Glass thermometers   NEVER use glass thermometers to check food temperatures. If they break, they can be a physical hazard.

Checking temperatures   When checking the temperature of food, insert the probe into the thickest part of the food, as shown in the photo at left. This is usually in the center. Also take another reading in a different spot. The temperature may vary in different areas.

Before recording a temperature, wait for the thermometer reading to steady. Wait at least 15 seconds after you insert the stem or the probe.

Apply Your Knowledge

Pick the Right Thermometer

For each situation, choose the best thermometer or thermometers. Some thermometers may be chosen more than once. Write the letter or letters in the space provided.

① _____ Internal temperature of a chicken breast

② _____ Internal temperature of a roast

③ _____ Internal temperature of a large stockpot of soup

④ _____ Surface temperature of a grill

⑤ _____ Air temperature of a cooler

Ⓐ Bimetallic stemmed thermometer

Ⓑ Thermocouple with immersion probe

Ⓒ Thermocouple with surface probe

Ⓓ Thermocouple with penetration probe

Ⓔ Thermocouple with air probe

Ⓕ Infrared thermometer

Calibrate the Thermometer

Put the steps for calibrating a bimetallic stemmed thermometer in the right order by writing the number of the step in the space provided.

_____ Rotate the head of the thermometer until it reads 32°F (0°C).

_____ Put the thermometer stem into the ice water and wait 30 seconds.

_____ Fill a container with crushed ice and tap water.

_____ Hold the adjusting nut with a wrench or other tool.

For answers, please turn to page 5.16.

Chapter Summary

The flow of food is the path food takes in your operation from purchasing to service. Many things can happen to food in its flow through the operation. Two major concerns are cross-contamination and time-temperature abuse. To prevent cross-contamination, use separate equipment for each type of food. Also, you must clean and sanitize all work surfaces, equipment, and utensils after each task. Prepping ready-to-eat food before raw food is one way to minimize the chance for cross-contamination. Similarly, you can buy food items that don't require much preparation or handling. Time-temperature abuse happens any time food remains between 41°F and 135°F (5°C and 57°C). This range is called the temperature danger zone. You must try to keep food out of this range.

A thermometer is the most important tool you can use to prevent time-temperature abuse. You should regularly record food temperatures and the times they were taken. Always put the thermometer stem or probe into the thickest part of the food. A bimetallic thermometer should be put in food from the tip to the end of the sensing area. Before you record the temperature, wait for the thermometer reading to steady. Never use glass thermometers with food items. Thermometers can be calibrated in two ways: the boiling-point method or the ice-point method. Thermometers should be calibrated regularly. Most important, they must be cleaned and sanitized before and after each use.

Chapter Review Case Study

Now take what you have learned in this chapter and apply it to the following case study.

At 6:00 a.m., Kim started her workday at The Little Bistro. After a quick meeting with the chef, Kim knew what she would be doing that day. Her first task was to make the broccoli quiches for the lunch special. By 6:15 a.m., she had already collected all the ingredients except the broccoli. She brought out salt, eggs, cream, butter, and cheese and set them on the prep table by the mixer. On her last trip to the cooler, she got the broccoli. It took over an hour to wash and chop it. Finally, she was able to mix the quiche filling. She started by cracking the eggs in a bowl. Then she added the remaining ingredients. She dripped egg whites on the table, making a mess. Leaving the leftover eggs and cream on the table, she quickly got out the premade quiche crusts from the freezer and poured the filling. By the time she got the quiches in the oven, it was 10:30 a.m.

Kim hurried to start her next task—making fruit salad. She quickly removed the rinds from some melons and then sliced them on the salad prep table. The melon juice made a mess, so she wiped down the table. Before she could start on the other fruit, the oven timer went off. Kim needed to check the quiches. They were supposed to bake for around 30 minutes. However, she did not want to overcook them. The chef said their internal temperature needed to be 155°F (68°C). She used an infrared thermometer to check the temperature of one quiche in two places. The readings were both in the right range. She took the quiches out of the oven and set them on a table to cool. While they cooled, Kim went back to the fruit salad. It was already 11:00 a.m., and the salad wasn't ready. She hurried to prep the strawberries, kiwi, and grapes on the same table as the melons. When she headed back to the cooler to get the citrus dressing for the salad, she noticed the eggs and cream she had left out.

At 11:45 a.m., the lunch rush was in full swing. She grabbed the eggs on her way back to the cooler and put them away. After adding the dressing to the salad, she put the salad and the dressing in the cooler. Then she wiped down the prep table. Knowing that they needed the table by the mixer, she put away the leftover cream. After the rush slowed down, Kim cleaned and sanitized the mixer and both of the tables she used.

What did Kim do wrong?

_____

_____

_____

For answers, please turn to page 5.16.

Study Questions

Circle the best answer to each question below.

① **A foodhandler has finished trimming raw chicken on a cutting board and needs it to prep vegetables. What must be done to the cutting board?**

  A It must be dried with a paper towel.

  B It must be turned over to the other side.

  C It must be washed, rinsed, and sanitized.

  D It must be rinsed in hot water and air-dried.

② **Which of these practices can help prevent cross-contamination?**

  A Using a designated cutting board when prepping meat

  B Preparing small batches of food at one time

  C Identifying minimum internal cooking temperatures

  D Calibrating thermometers in the operation regularly

③ **Infrared thermometers should be used to measure the**

  A air temperature in a cooler.

  B internal temperature of a turkey.

  C surface temperature of a grill.

  D internal temperature of a batch of soup.

④ **At what temperatures do foodborne pathogens grow most quickly?**

  A Between 0°F and 41°F (–17°C and 5°C)

  B Between 45°F and 65°F (7°C and 18°C)

  C Between 70°F and 125°F (21°C and 52°C)

  D Between 130°F and 165°F (54°C and 74°C)

⑤ **Which thermocouple probe should be used to check the temperature of a large stockpot of soup?**

  A Air probe

  B Surface probe

  C Immersion probe

  D Penetration probe

⑥ **When a thermometer is calibrated using the ice-point method, it should be adjusted to _____ after the stem or probe has been placed in the ice water.**

A 0°F (−17°C)

B 32°F (0°C)

C 41°F (5°C)

D 212°F (100°C)

⑦ **What type of thermometer is NOT appropriate for use in a restaurant or foodservice operation?**

A Thermistor

B Thermocouple

C Glass thermometer

D Bimetallic stemmed thermometer

For answers, please turn to page 5.16.

Answers

5.5   An Ounce of Prevention

1 and 3 should be marked.

5.5   Is It Safe?

① No. Anita took out more tuna salad than she needed to make a small number of sandwiches. This exposed the tuna salad to time-temperature abuse, which was made worse by the many interruptions.

② No. He did not sanitize the table and equipment after he cut up the chickens. The onions and peppers could have been contaminated by the chickens.

5.12   Pick the Right Thermometer

① A, D   ④ C, F

② A, D   ⑤ E

③ B

5.12   Calibrate the Thermometer

4, 2, 1, 3

5.13   Chapter Review Case Study

Here is what Kim did wrong.

- She left the eggs and dairy at room temperature for too long. The quiche filling was at room temperature for four hours and 15 minutes. The leftover eggs and dairy were at room temperature for five and a half hours. She should have thrown away the leftover egg and dairy.

- She used the wrong kind of thermometer to check the internal temperature of the quiches.

- She let the quiches cool at room temperature and did not store them correctly.

- She did not clean and sanitize the prep table by the mixer after she finished preparing the quiches.

- She did not clean and sanitize the salad prep table after she cut the melons and before prepping the other fruit.

- Kim did not wash her hands at all. She should have washed her hands before making the quiches, before prepping the melons, and before prepping the other fruit.

5.14   Study Questions

① C   ⑤ C

② A   ⑥ B

③ C   ⑦ C

④ C

Notes

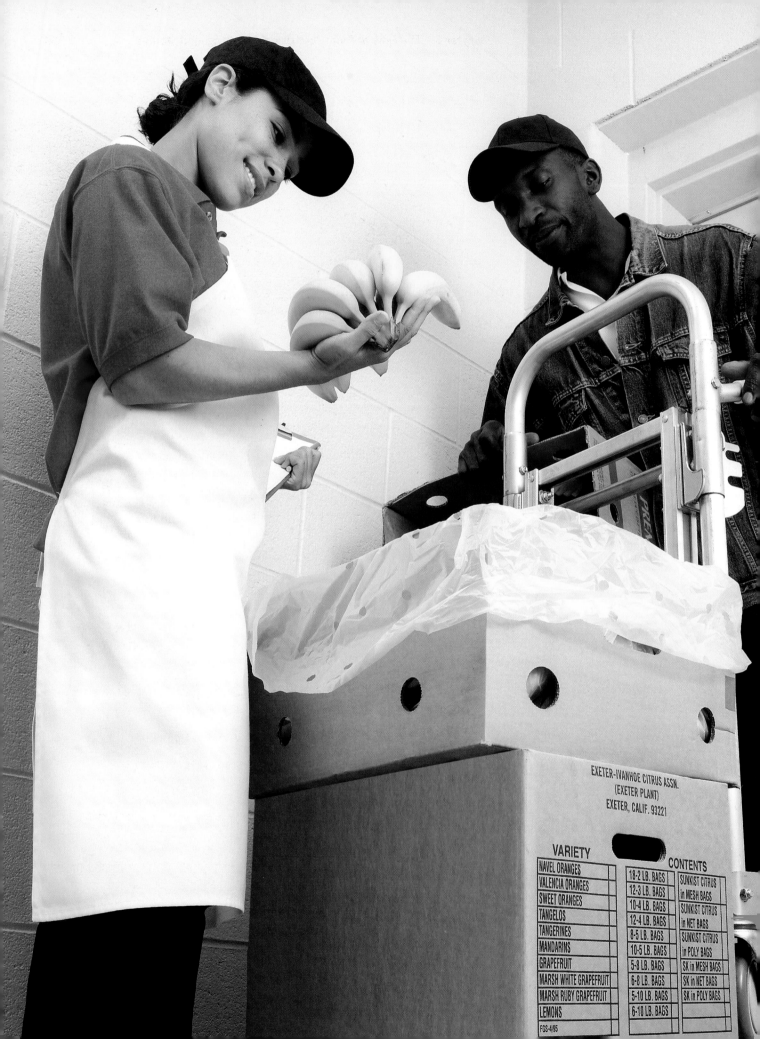

EXETER-IVANHOE CITRUS ASSN.
(EXETER PLANT)
EXETER, CALIF. 93221

VARIETY		CONTENTS
NAVEL ORANGES	18-2 LB. BAGS	SUNKIST CITRUS in MESH BAGS
VALENCIA ORANGES	12-3 LB. BAGS	
SWEET ORANGES	10-4 LB. BAGS	SUNKIST CITRUS in NET BAGS
TANGELOS	12-4 LB. BAGS	
TANGERINES	8-5 LB. BAGS	SUNKIST CITRUS in POLY BAGS
MANDARINS	10-5 LB. BAGS	
GRAPEFRUIT	5-8 LB. BAGS	SK in MESH BAGS
MARSH WHITE GRAPEFRUIT	6-8 LB. BAGS	SK in NET BAGS
MARSH RUBY GRAPEFRUIT	5-10 LB. BAGS	SK in POLY BAGS
LEMONS	6-10 LB. BAGS	

FGS-4/95

6

The Flow of Food: Purchasing, Receiving, and Storage

In the News

A Costly Finishing Touch

A local steakhouse recently avoided a possible crisis thanks to one of its managers. On a routine self-inspection at the restaurant, the manager saw cooks on the grill line using an unlabeled spray bottle of olive oil to finish some steak and chicken dishes.

Later in the lunch rush, a cook plated steaks and sprayed them with what she thought was olive oil. The manager noticed that the cook was using a different spray bottle than before. Upon examining the steaks before they were served, the manager discovered that what the cook had thought to be olive oil was actually degreaser.

The cook admitted that she kept a spray bottle of degreaser handy to clean the grill. The cook had mistaken the degreaser for olive oil, because both liquids had been stored in unlabled spray bottles. The manager solved the problem by immediately throwing out the steaks and then updating policies so that all spray bottles are labeled and degreasers are stored away from prep areas.

You Can Prevent This

In the story above, incorrect labeling and incorrect storing led to contaminated food. Fortunately, the problem was found before customers became sick.

In this chapter, you will learn the following guidelines for keeping food safe during purchasing, receiving, and storage.

- Purchasing food from approved, reputable suppliers

- Using criteria to accept or reject food during receiving

- Labeling and dating food

- Storing food and nonfood items to prevent time-temperature abuse and contamination

---

Concepts from Earlier Chapters

Before reading this chapter, remember these concepts and facts.

Temperature danger zone   Temperature range between 41°F and 135°F (5°C to 57°C). Foodborne pathogens grow well in it.

Time-temperature abuse   When food stays too long at temperatures that are good for pathogen growth.

General Purchasing and Receiving Principles

You can't make unsafe food safe. So, you must make sure you bring only safe food into your operation. Purchasing food from approved, reputable suppliers and following good receiving procedures will help to ensure the safety and quality of the food your operation uses.

Purchasing

Before you accept any deliveries, you must make sure that the food you purchase is safe. It should come from approved, reputable suppliers. Follow these guidelines.

Photo courtesy of Boskovich Farms, Inc.

Approved, reputable suppliers   An approved food supplier is one that has been inspected and meets all applicable local, state, and federal laws. Make sure your suppliers have good food safety practices. This applies to all suppliers along the supply chain. Your operation's supply chain can include growers (as shown in the photo at left), shippers, packers, manufacturers, distributors (trucking fleets and warehouses), or local markets.

Develop a relationship with your suppliers, and get to know their food safety practices. Consider reviewing their most recent inspection reports. These reports can be from the U.S. Department of Agriculture (USDA), the Food and Drug Administration (FDA), or a third-party inspector. They should be based on Good Manufacturing Practices (GMP) or Good Agricultural Practices (GAP). Make sure the inspection report reviews the following areas.

- Receiving and storage
- Processing
- Shipping
- Cleaning and sanitizing
- Personal hygiene
- Employee training
- Recall program
- HACCP program or other food safety system

Deliveries   Arrange deliveries so they arrive one at a time and during off-peak hours. Suppliers must deliver products when staff has enough time to do inspections.

Receiving and Inspecting

You must take steps to ensure the receiving and inspection process goes smoothly and keeps the food safe. Make sure enough trained staff are available to promptly receive, inspect, and store food. Authorize staff to accept, reject, and sign for deliveries. Deliveries should be carefully and immediately inspected and put away as quickly as possible. This is especially true for refrigerated and frozen items.

If you must reject an item, set it aside from the items you are accepting. Then tell the delivery person exactly what's wrong with the rejected item. Make sure you get a signed adjustment or credit slip before the rejected item is thrown out or given back to the delivery person. Finally, log the incident on the invoice or the receiving document.

Temperature

Use thermometers to check food temperatures during receiving. The table below shows how to check the temperature of various food.

Checking the Temperature of Various Types of Food

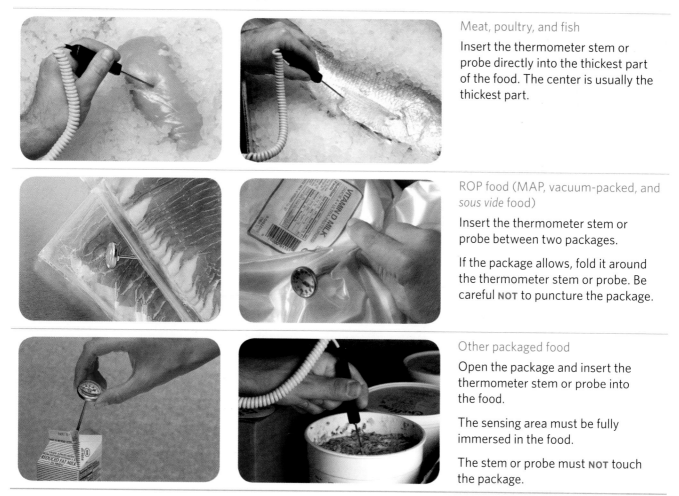

Meat, poultry, and fish

Insert the thermometer stem or probe directly into the thickest part of the food. The center is usually the thickest part.

ROP food (MAP, vacuum-packed, and *sous vide* food)

Insert the thermometer stem or probe between two packages.

If the package allows, fold it around the thermometer stem or probe. Be careful **NOT** to puncture the package.

Other packaged food

Open the package and insert the thermometer stem or probe into the food.

The sensing area must be fully immersed in the food.

The stem or probe must **NOT** touch the package.

Deliveries should also meet the following temperature criteria.

Cold food   Receive cold TCS food at 41°F (5°C) or lower, unless otherwise specified.

Hot food   Receive hot TCS food at 135°F (57°C) or higher.

Frozen food   Receive frozen food frozen. Reject frozen food for the following reasons.

- Fluids or frozen liquids appear in case bottoms.
- There are ice crystals on the product or the packaging. Or there are water stains on the packaging. This may be evidence of thawing and refreezing. The food in the photo at left shows evidence of thawing and refreezing.

Packaging

The packaging of food and nonfood items should be intact and clean. It should protect food and food-contact surfaces from contamination. Reject food and nonfood items if packaging has any of the following problems.

Damage   Reject items with tears, holes, or punctures in their packaging. Likewise, reject cans with swollen ends, rust, or dents. Items with broken cartons or seals or dirty wrappers should also be rejected.

Liquid   Reject items with leaks, dampness, or water stains (which means the item was wet at some point), as shown in the photo at left.

Pests   Reject items with signs of pests or pest damage.

Dates   Reject items with expired code or use-by dates.

Product Quality

Poor food quality can be a sign that the food has been time-temperature abused and, therefore, may be unsafe. Work with your suppliers to define specific safety and quality criteria for the products you typically receive. Reject food if it has any of the following problems.

Color   Reject food with an abnormal color.

Texture   Reject meat, fish, or poultry that is slimy, sticky, or dry. Also reject it if it has soft flesh that leaves an imprint when you touch it.

Odor   Reject food with an abnormal or unpleasant odor.

In addition to the guidelines above, you should always reject any item that does not meet your company's standards for quality.

Receiving and Inspecting Specific Food

You must follow the general receiving and inspection guidelines for all food and nonfood items your operation receives. In addition, some types of food require specific guidelines.

Eggs

Eggs must be clean and unbroken when you receive them. The eggs in the photo at the left must be rejected. Reject eggs if they do not meet the following guidelines.

- Shell eggs must be received at an air temperature of 45°F (7°C) or lower.

- Liquid, frozen, and dehydrated egg products must be pasteurized as required by law and have a USDA inspection mark. Eggs also must comply with USDA grade standards. Refer to page 6.6 for a picture of the inspection mark.

Milk and Dairy Products

Milk and dairy products must be received at 41°F (5°C) or lower unless otherwise specified by law. They also must be pasteurized and comply with FDA grade A standards.

Shellfish

Shellfish can be received either shucked or live.

Raw shucked shellfish   Make sure that raw shucked shellfish are packaged in nonreturnable containers.

- Containers must be labeled with the packer's name, address, and certification number.

- Containers smaller than one-half gallon must have either a "best if used by" or "sell by" date. Containers bigger than one-half gallon (1.9 L) must have the date the shellfish were shucked.

Live shellfish   Receive with shellstock identification tags.

- These tags, as shown in the photo at left, must remain attached to the delivery container until all of the shellfish have been used. Employees must write on the tags the date that the last shellfish was sold or served from the container. Operators must keep these tags on file for 90 days from the date written on them.

- Reject shellfish if they are very muddy, have broken shells, or are dead.

Produce

Sliced melons, cut tomatoes, and fresh-cut leafy greens must be received at 41°F (5°C) or lower.

Prepackaged Juice

Prepackaged juice must be purchased from a supplier with a Hazard Analysis Critical Control Point (HACCP) plan. (You will learn about HACCP plans in chapter 9.) The juice must be treated (e.g., pasteurized) to prevent, eliminate, or reduce pathogens.

Fish Served Raw or Partially Cooked

The supplier must freeze fish that will be served raw or partially cooked, such as sushi-grade fish, for a specific period of time to kill any parasites that might be in the fish. The supplier should freeze fish to one of the following temperatures prior to shipment.

- −4°F (−20°C) or lower for at least seven days (168 hours) in a freezer

- −31°F (−35°C) or lower until frozen solid and then stored at −31°F (−35°C) for at least 15 hours

- −31°F (−35°C) or lower until frozen solid and then stored at −4°F (−20°C) or lower for at least 24 hours

Your supplier will provide you with records showing that the fish was frozen correctly. You must keep these records on file for 90 days from the date you served the fish.

Products Requiring Inspection Stamps

Checking for inspection stamps is a way to make sure food is coming from an approved source. The inspection stamps for meat and poultry and for egg products are shown below.

Meat and poultry

Packaging must have a USDA or a state department of agriculture's inspection stamp. This stamp indicates that the product and the processing plant have met certain standards. The stamp at left is for meat.

Egg products

Packaging must have an inspection stamp indicating that federal regulations have been enforced to maintain quality and reduce contamination.

Apply Your Knowledge

Accept or Reject?

Place an A next to the food items you should accept. Place an R next to the food items you should reject.

① _____ Chicken received at an internal temperature of 50°F (10°C)

② _____ Can of red kidney beans with a small dent on one side of the can

③ _____ Eggs received at an air temperature of 45°F (7°C)

④ _____ Fresh salmon with flesh that springs back when touched

⑤ _____ Bag of flour that is dry but has a watermark on it

⑥ _____ Live oysters without shellstock identification tags

⑦ _____ Frozen meat with ice crystals on the packaging

⑧ _____ Milk received at 50°F (10°C)

⑨ _____ Sushi-grade tuna frozen until solid at 20°F (–7°C) for 72 hours

⑩ _____ Vacuum-packed bacon with the seal broken but no other obvious damage

For answers, please turn to page 6.18.

Storing

Following good storage guidelines for food and nonfood items will help keep these items safe and preserve their quality. In general, you must take steps to keep your storage areas in good condition and rotate your stock. More specifically, you also must follow some guidelines for storing refrigerated, frozen, and dry food.

General Storage Guidelines

Use the following general guidelines when storing food.

PATHOGEN PREVENTION

Labeling

- Label all TCS, ready-to-eat food prepped in-house that you have held for longer than 24 hours. The label, as shown in the photo at left, must include the name of the food and the date by which it should be sold, eaten, or thrown out.

  You can store all ready-to-eat, TCS food that has been prepped in-house for a maximum of seven days at 41°F (5°C) or lower. After seven days, you must throw it out to prevent bacteria, such as *Listeria monocytogenes,* from growing to unsafe levels.

- Label food prepped in-house that has been made with previously cooked and stored food with the discard date of the previously cooked item. For example, if you use previously cooked ground beef to make meat sauce, you must label the meat sauce with the discard date of the ground beef.

Rotation

- Rotate food to use the oldest inventory first. Many operations use the first-in, first-out (FIFO) method to rotate their refrigerated, frozen, and dry food during storage. Here is one way to use the FIFO method.

  1 Identify the food item's use-by or expiration date.

  2 Store items with the earliest use-by or expiration dates in front of items with later dates, as shown in the photo at left.

  3 Once shelved, use those items stored in front first.

- Make a schedule to throw out stored food on a regular basis. If your operation has not sold or used a food item by a specified date, throw it out. Clean and sanitize the container and refill it with fresh food. For example, use flour stored in plastic bins, as shown in the photo at left, within 12 months from the time you placed it in the bins. After 12 months, throw out the remaining flour and clean and sanitize the bins. Then refill the bins with new flour.

- Throw away food that has passed its manufacturer's use-by or expiration date.

Temperatures

- Keep TCS food at 41°F (5°C) or lower, or at 135°F (57°C) or higher.

- Check the temperature of stored food and storage areas at the beginning of the shift. Many operations use a preshift checklist to guide staff through this process.

Product Placement

- Store food in containers intended for food. The containers should be durable, leak proof, and able to be sealed or covered. NEVER use empty food containers to store chemicals. NEVER put food in empty chemical containers.

- Store food, linens, and single-use items in designated storage areas.

- Store food, linens, and single-use items away from walls and at least six inches (15 centimeters) off the floor.

Cleaning

- Keep all storage areas clean and dry. Clean floors, walls, and shelving in coolers, freezers, dry-storage areas, and heated holding cabinets on a regular basis, as shown in the photo at left. Clean up spills and leaks right away to keep them from contaminating other food.

- Clean dollies, carts, transporters, and trays often.

- Store food in containers that have been cleaned and sanitized.

- Store dirty linens in a clean, washable container in a way that prevents the contamination of food.

Refrigerated and Frozen Storage

As a manager, you are responsible for making sure that your operation's coolers and freezers are keeping cold food cold and frozen food frozen. When coolers and freezers are not working the right way, the food in them can become time-temperature abused. Follow the guidelines below for refrigerated and frozen storage.

Maintenance

- Schedule regular maintenance, as shown in the photo at left, on coolers and freezers to make sure they stay at the right temperatures.

- Defrost freezers to allow them to operate more efficiently.

Temperatures

- Set the temperature of coolers to keep the internal temperature of TCS food at 41°F (5°C) or lower, unless otherwise indicated by the manufacturer or your regulatory authority. Fresh eggs must be kept at an air temperature of 45°F (7°C) or lower.

- Set freezers to keep products frozen. This temperature will vary from product to product.

Monitoring

- Monitor food temperatures regularly. Randomly sample the temperature of stored food to verify that the cooler is working.

- Check cooler temperatures at least once during each shift. Place hanging thermometers inside the cooler to make this task easy to do. Some coolers have a temperature readout on the outside, as shown in the photo at left. Also check these for accuracy.

Airflow

- Do NOT overload coolers or freezers. Storing too many food items prevents good airflow and makes the units work harder to stay cold.

- Be aware that frequent opening of the cooler lets warm air inside, which can affect food safety. Consider using cold curtains in walk-in coolers to help maintain temperatures.

- Use open shelving. Lining shelves with aluminum foil, sheet pans, or paper restricts circulation of cold air in the unit.

Preventing Cross-Contamination

- Store food in ways that prevent cross-contamination. Wrap or cover food. Store refrigerated raw meat, poultry, and seafood separately from ready-to-eat food. If raw and ready-to-eat food cannot be stored separately, store ready-to-eat food above raw meat, poultry, and seafood, as shown in the photo below. This will prevent juices from raw food from dripping onto ready-to-eat food. Raw meat, poultry, and seafood can be stored with or above ready-to-eat food in a freezer if all of the items have been commercially processed and packaged.

- Store raw meat, poultry, and seafood in coolers in the following top-to-bottom order: seafood, whole cuts of beef and pork, ground meat and ground fish, whole and ground poultry. This order is based on the minimum internal cooking temperature of each food.

A Ready-to-eat food

B Seafood

C Whole cuts of beef and pork

D Ground meat and ground fish

E Whole and ground poultry

Dry Storage

Follow these guidelines when placing food and other items in dry storage.

- Keep dry-storage areas cool and dry. To keep food at its highest quality and to assure food safety, the temperature of the dry-storage area should be between 50°F and 70°F (10°C and 21°C).

- Store dry food away from walls and at least six inches (fifteen centimeters) off the floor, as shown in the photo at left.

- Make sure dry-storage areas are well ventilated to help keep temperature and humidity constant throughout the storage area.

6" (15 cm)

Apply Your Knowledge

Load the Cooler

Next to the number of each food item, write the letter of the shelf it belongs on.

① _____ Whole meat

② _____ Whole chicken

③ _____ Pecan pie

④ _____ Raw ground beef

What's Wrong with This Picture?

Find the unsafe storage practices in this picture.

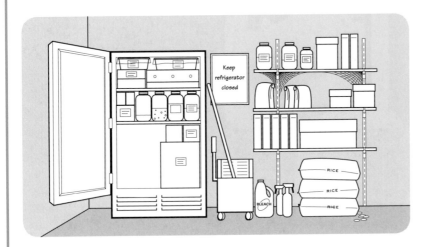

_____

_____

_____

_____

For answers, please turn to page 6.18.

Chapter Summary

Always purchase food from approved, reputable suppliers. Suppliers must be inspected regularly and be in compliance with local, state, and federal laws. Develop a relationship with your suppliers, and get to know their food safety practices.

Plan delivery schedules so that food is handled the right way. Staff should know how to accept or reject food during receiving. Food must be delivered at the right temperature. Frozen food should not have been thawed and refrozen. Staff should check the color, texture, and odor of all food, especially meat, poultry, and fish. The packaging of all deliveries should be clean and undamaged. Use-by dates should be current, and food should show no signs of mishandling.

Rotate food in storage to use the oldest inventory first. Label all TCS, ready-to-eat food prepped in-house that you have held for longer than 24 hours. Throw away food that has passed its expiration date. Store refrigerated raw meat, poultry, and seafood separately from ready-to-eat food. If you cannot store them separately, store ready-to-eat food above raw seafood, meat, and poultry.

Chapter Review Case Study

Now take what you have learned in this chapter and apply it to the following case study.

A shipment was delivered to Enrico's Italian Restaurant on a warm summer day. Alyce, who was in charge of receiving, began inspecting the shipment.

First, she inspected the bags of frozen shrimp. Alyce noticed the ice crystals inside the bags and took that as a good sign that the shrimp were still frozen. Next she used a thermometer to test the temperature of the vacuum-packed packages of ground beef, which was 40°F (4°C). Then Alyce used the same thermometer to measure the temperature of the fresh salmon. The salmon was on ice, although it seemed as though much of the ice had melted. The internal temperature of the salmon was 43°F (6°C), and the flesh sprung back after she touched it. She accepted the ground beef and the salmon and placed them on the side to put away.

Not wanting to take the time to clean and sanitize the probe, Alyce felt several containers of sour cream. They felt cold, so Alyce also placed them on the side to put away. Finally, Alyce inspected the cases of pasta. One of the cases was torn, but the pasta inside the case didn't seem to be damaged.

Once she finished receiving the food, Alyce was ready to put it into storage. First, she carried the bags of shrimp to the freezer. She wondered who had left the freezer without making sure the door was completely shut. Alyce then loaded a case of sour cream on the dolly and wheeled it over to the reach-in cooler.

When she opened the cooler, she noticed that it was tightly packed. However, she was able to squeeze the case into a spot on the top shelf. Next, she wheeled several cases of fresh ground beef and the fresh salmon over to the walk-in cooler. She noticed that the readout on the outside of the cooler indicated 39°F (4°C). Alyce pushed through the cold

Continued on the next page ▶

► *Continued from previous page*

curtains and bumped into a stockpot of soup as she moved inside. She moved the soup over and made a space next to the door for the ground beef. She was able to put the salmon on the shelf above the soup.

Alyce said hello to Mary, who had just cleaned the shelving in the unit and was lining it with new aluminum foil. Alyce returned to the receiving area and loaded several cases of pasta on the dolly. She was sweating as she stacked the boxes on the shelf in dry storage and gave a quick glance at the thermometer in the dry-storage room, which read 85°F (29°C). When she was finished stacking the boxes, Alyce returned the dolly to the receiving area.

① What receiving mistakes did Alyce make?

_____

_____

_____

_____

② What storage mistakes were made at the operation?

_____

_____

_____

_____

For answers, please turn to page 6.18.

Study Questions

Circle the best answer to each question below.

① **What is the most important factor in choosing an approved food supplier?**

A It has a HACCP program or other food safety system.

B It has documented manufacturing and packing practices.

C Its warehouse is close to the establishment, reducing shipping time.

D It has been inspected and complies with local, state, and federal laws.

② **Containers of raw shucked shellfish bigger than one-half gallon (1.9 L) must have the name and address of the packer, the certification number, and a**

A shellstock identification tag.

B shucked date.

C harvest date.

D USDA inspection mark.

③ **What is the maximum acceptable receiving temperature for fresh beef?**

A 35°F (2°C)

B 41°F (5°C)

C 45°F (7°C)

D 50°F (10°C)

④ **In top-to-bottom order, how should a fresh pork roast, fresh salmon, a carton of lettuce, and a pan of fresh chicken breasts be stored in a cooler?**

A Lettuce, fresh salmon, fresh pork roast, fresh chicken breasts

B Fresh salmon, fresh pork roast, fresh chicken breasts, lettuce

C Lettuce, fresh chicken breasts, fresh pork roast, fresh salmon

D Fresh salmon, lettuce, fresh chicken breasts, fresh pork roast

⑤ **What is the warmest acceptable receiving temperature for eggs?**

A 32°F (0°C)

B 41°F (5°C)

C 45°F (7°C)

D 50°F (10°C)

⑥ **Where should raw poultry be placed in a cooler?**

A On the top shelf

B Next to produce

C On the bottom shelf

D Above ready-to-eat food

Continued on the next page ▶

► *Continued from previous page*

⑦ **Why is first-in, first-out (FIFO) storage used?**

A To ensure that the oldest food is used first

B To ensure that the newest food is used first

C To reduce the time it takes to store new shipments

D To use food that has passed its expiration date

⑧ **When storing TCS, ready-to-eat food that was prepared on-site, what information must be included on the label?**

A Potential allergens

B Product ingredients

C Nutritional information

D Sell-by or discard date

⑨ **What is the warmest temperature at which ground beef can be safely stored?**

A 0°F (-17°C)

B 32°F (0°C)

C 41°F (5°C)

D 60°F (16°C)

⑩ **Large ice crystals in a case of frozen food are evidence that the product may have been**

A received at 6°F to 10°F (-14°C to -12°C).

B stored at 6°F to 10°F (-14°C to -12°C).

C shipped correctly.

D thawed and refrozen.

⑪ **How should cartons of coleslaw be checked for correct receiving temperature?**

A Check the interior air temperature of the delivery truck.

B Open a carton and insert a thermometer stem into the food.

C Place a thermometer against the outside of the carton.

D Touch the carton to see if it is cold.

⑫ **A box of sirloin steaks carries a state department of agriculture inspection stamp. What does this stamp indicate?**

A The steaks are free of disease-causing microorganisms.

B The meat and processing plant have met certain standards.

C The meat wholesaler meets USDA or a state department of agriculture's quality-grading standards.

D The farm that supplied the beef uses only certified animal feed.

⑬ **At what temperature should dry-storage rooms be kept?**

A 35°F to 40°F (2°C to 5°C)

B 40°F to 60°F (5°C to 16°C)

C 50°F to 70°F (10°C to 21°C)

D 75°F to 90°F (24°C to 32°C)

⑭ **When storing food using the FIFO method, the food with the earliest use-by dates should be stored**

A below food with later use-by dates.

B behind food with later use-by dates.

C in front of food with later use-by dates.

D alongside food with later use-by dates.

⑮ **An operation that has prepped tuna salad can store it at 41°F (5°C) or lower for a maximum of how many days?**

A 1

B 3

C 7

D 14

⑯ **It is important to avoid lining cooler shelves with aluminum foil because the foil**

A can restrict the flow of cold air.

B can give food a metallic flavor.

C reduces visibility.

D prevents leaks from reaching the floor drain.

For answers, please turn to page 6.19.

Answers

6.7   Accept or Reject?

① R   ⑥ R

② R   ⑦ R

③ A   ⑧ R

④ A   ⑨ R

⑤ R   ⑩ R

6.12   Load the Cooler

① B   ③ A

② D   ④ C

6.12   What's Wrong with This Picture?

Here are the unsafe storage practices.

- Chemicals stored with food

- Food stored on the floor

- Boxes of food not labeled

- Spilled food not cleaned up

- Cooler door open

- Overstocked cooler

- Ready-to-eat food on the wrong shelf

- Area not clean

- Unlabeled items in cooler

6.13   Chapter Review Case Study

① Alyce made the following receiving mistakes.

- She should have rejected the shrimp. The ice crystals are evidence of thawing and refreezing.

- She did not clean and sanitize the probe she had used to measure the temperature of the ground beef and the fish.

- She should have rejected the salmon. The temperature of the fish was above 41°F (5°C), and the melted ice could be evidence of time-temperature abuse.

- She felt the container of sour cream instead of measuring the internal temperature of the food.

- She should have rejected the torn carton of pasta.

- She put the cold food on the side while receiving the dry food.

② The operation made the following storage mistakes.

- The freezer door was left open.

- Alyce placed the case of sour cream into an already overloaded cooler.

- Alyce put the raw salmon above ready-to-eat food (soup).

- Alyce checked the cooler's readout temperature instead of spot-checking the internal temperatures of the food stored inside.

- A stockpot of soup was stored on the cooler floor.

- Mary was lining the cooler shelving with aluminum foil. This can restrict airflow in the unit.

- The temperature in the dry-storage room was 85°F (29°C), which is too warm. Dry-storage areas should be between 50°F and 70°F (10°C and 21°C).

6.15   Study Questions

① D   ⑨ C

② B   ⑩ D

③ B   ⑪ B

④ A   ⑫ B

⑤ C   ⑬ C

⑥ C   ⑭ C

⑦ A   ⑮ C

⑧ D   ⑯ A

7

The Flow of Food: Preparation

In the News

Cross-Country Outbreak

An elderly woman died and several hundred others became sick after eating at a dinner held at a state fair. County health officials said the source was stuffed ham served at the dinner.

The woman who died had eaten the meal after it was delivered to her home. Most of the others who became sick were tourists. Within days of the dinner, regulatory authorities across the country received reports of illness. Those who were hospitalized experienced nausea and vomiting. They were later diagnosed with staphylococcal gastroenteritis.

Health officials believed that the stuffed ham caused the illness because it was undercooked. It was also stacked in a walk-in cooler to cool. They believed this allowed *Staphylococcus aureus* to grow.

You Can Prevent This

The foodborne-illness outbreak at the state fair happened because the stuffed ham wasn't cooked or cooled correctly. You can prevent a situation like this if you know how to avoid time-temperature abuse and cross-contamination when prepping TCS food.

In this chapter, you will learn the following guidelines for keeping food safe during preparation.

- Thawing food correctly

- Preventing cross-contamination and time-temperature abuse

- Cooking food to a minimum internal temperature

- Cooling and reheating food to the right temperature in the right amount of time

---

Concepts from Earlier Chapters

Before reading this chapter, remember these concepts and facts.

Temperature danger zone   Temperature range between 41°F and 135°F (5°C to 57°C). Foodborne pathogens grow well in it.

When to wash your hands   Staff must wash hands after using the restroom or touching something that may contaminate their hands.

Correct glove use   Gloves should be used only after washing hands and must be changed often.

Thermometer calibration and use   Calibrate thermometers often, using either the ice-point or the boiling-point method.

Preparation

You have purchased, received, and stored your operation's food correctly. Now you must prepare it. Cross-contamination and time-temperature abuse can happen easily during this step in the flow of food. But you can prevent pathogens from growing during preparation by making good food-prep choices. It is important to use the right thawing methods. You also need to be aware of special handling practices when prepping specific kinds of food. Some practices will also require a variance from your local regulatory authority.

General Preparation Practices

No matter what type of food you are prepping, you should begin by following these guidelines.

Equipment   Make sure workstations, cutting boards, and utensils are clean and sanitized.

Quantity   Remove from the cooler only as much food as you can prep in a short period of time. Also prep food in small batches. This keeps ingredients from sitting out for long periods of time. In the photo at left, the foodhandler has taken out too much tuna salad.

Storage   Return prepped food to the cooler, or cook it as quickly as possible.

Thawing

Freezing does not kill pathogens. When frozen food is thawed and exposed to the temperature danger zone, any pathogens in the food will begin to grow. To reduce this growth, NEVER thaw food at room temperature. You must thaw TCS food in one of these ways.

Refrigeration   Thaw food in a cooler, at a product temperature of 41°F (5°C) or lower.

Running water   Submerge food under running water at 70°F (21°C) or lower. The photo at left shows the right way to do this. Make sure the water is potable—safe to drink.

Microwave   Thaw food in a microwave oven if it will be cooked just after thawing.

Cooking   Thaw food as part of the cooking process.

Prepping Specific Food

Some food requires special care during preparation. Produce, batter and breading, eggs and egg mixtures, and salads that contain TCS food all need extra consideration. Even ice needs special handling.

Produce

When prepping produce, follow these guidelines.

Cross-contamination   Make sure fruit and vegetables do **NOT** touch surfaces exposed to raw meat or poultry.

Washing   Wash fruit and vegetables thoroughly under running water. Do this before cutting, cooking, or combining it with other ingredients.

- The water should be a little warmer than the produce.

- Pay special attention to leafy greens such as lettuce and spinach, as the foodhandler in the photo at left is doing. Remove the outer leaves, and pull the lettuce or spinach completely apart and rinse thoroughly.

Sanitizing   Produce can be sanitized by washing it in water containing ozone. Check with your local regulatory authority to see if this is allowed in your area.

Soaking or storing   When soaking or storing produce in standing water or an ice-water slurry, do **NOT** mix different items or multiple batches of the same item.

Fresh-cut produce   Refrigerate and hold sliced melons, cut tomatoes, and cut leafy greens at 41°F (5°C) or lower. Consider holding other fresh-cut produce at this temperature.

Raw seed sprouts   If your operation primarily serves high-risk populations, do **NOT** serve raw seed sprouts.

Batter and Breading

Batters made with eggs or milk run the risk of time-temperature abuse and cross-contamination. Breading must also be handled with care, since cross-contamination is a risk. If you make breaded or battered food from scratch, follow these guidelines.

Batch size   Prep batter in small batches. Store what you don't need at 41°F (5°C) or lower in a covered container.

Discarding   Create a plan to throw out any unused batter or breading after at set amount of time, as shown in the photo at left. This might be after using a batch or at the end of a shift.

Allergens   Do **NOT** use the same batter or breading for different types of food if one of the food items can cause an allergic reaction.

Eggs and Egg Mixtures

When prepping eggs and egg mixtures, follow these guidelines.

Pooled eggs   Handle pooled eggs (if allowed by your local regulatory authority) carefully. Pooled eggs are eggs that are cracked open and combined in a container, as shown in the photo at left. Cook them promptly after mixing, or store them at 41°F (5°C) or lower. Wash and sanitize the containers used to hold them before making a new batch.

Pasteurized eggs   Consider using pasteurized shell eggs or egg products when prepping egg dishes that need little or no cooking. Examples include Caesar salad dressing, hollandaise sauce, tiramisu, and mousse.

High-risk populations   If you mainly serve high-risk populations, such as those in hospitals and nursing homes, use pasteurized eggs or egg products when serving dishes that are raw or undercooked. Shell eggs that are pooled must also be pasteurized. You may use unpasteurized shell eggs if the dish will be cooked all the way through, such as an omelet or a cake.

Salads Containing TCS Food

PATHOGEN PREVENTION

Chicken, tuna, egg, pasta, and potato salads all have been involved in foodborne-illness outbreaks. These salads usually are not cooked after preparation. This means you do not have a chance to get rid of pathogens, like hepatitis A, that may have gotten into the salad when it was made. Therefore, you must take a few extra steps. Follow these guidelines.

Using leftovers   Make sure leftover TCS food that will be used to make salads has been handled the right way. Food such as pasta, chicken, and potatoes should be used only if it has been cooked, held, and cooled correctly.

Storing leftovers   Throw out leftover food held at 41°F (5°C) or lower after seven days. Check the use-by date before using stored food items.

Fresh Juice Packaged On-Site

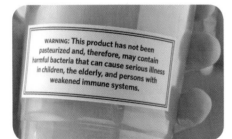

If you package fresh fruit or vegetable juice on-site for sale at a later time, you must treat (e.g., pasteurize) the juice according to an approved Hazard Analysis Critical Control Point (HACCP) plan.

If you don't treat the juice, it must be labeled as specified by federal regulation. An example is shown in the photo at left.

Ice

Ice has many uses in an operation. Follow these guidelines to avoid contaminating it.

Consumption   Make ice from water that is safe to drink.

Cooling food   **NEVER** use ice as an ingredient if it was used to keep food cold.

Containers and scoops   Use clean, sanitized containers and ice scoops to transfer ice from an ice machine to other containers.

- Store ice scoops outside of the ice machine in a clean, protected location, as shown in the photo at left.

- **NEVER** hold or carry ice in containers that have held raw meat, seafood, or poultry or chemicals.

- **NEVER** use a glass to scoop ice or touch ice with hands.

Preparation Practices That Require a Variance

You must get a variance when prepping food in certain ways. A variance is a document issued by your regulatory authority that allows a requirement to be waived or changed.

When applying for a variance, your regulatory authority may require you to submit a HACCP plan. The plan must account for any food safety risks related to the way you plan to prep the food item.

You need a variance if your operation plans to prep food in any of the following ways.

- Smoking food as a way to preserve it (but not to enhance flavor), as shown in the photo at left.

- Using food additives or adding components such as vinegar to preserve or alter the food so it no longer needs time and temperature control for safety.

- Curing food.

- Custom-processing animals. For example, this may include dressing deer in the operation for personal use.

- Packaging food using a reduced-oxygen packaging (ROP) method. This includes MAP, vacuum-packed, and *sous vide* food. *Clostridium botulinum* and *Listeria monocytogenes* are risks to food packaged in these ways.

- Sprouting seeds or beans.

- Offering live, molluscan shellfish from a display tank.

Apply Your Knowledge

What's the Problem?

Decide if the food was prepped or held correctly in each of the following examples. Then explain why or why not.

① Reggie filled a clean and sanitized sink with cold water and ice. Then he soaked two cases of green onions in it.

Was the food prepped correctly? _____

Why or why not? _____

_____

② Linda needed to make 20 box lunches to be picked up in three hours. She got out the bread, meat, and cheese and left them on the prep table, so she could make the sandwiches in between her other tasks.

Was the food prepped correctly? _____

Why or why not? _____

_____

③ Brandon trimmed an uncooked roast on the red cutting board. Then he washed his hands and used a different knife to slice tomatoes on the green cutting board.

Was the food prepped correctly? _____

Why or why not? _____

_____

④ Jessica read an article about sprouting beans. It inspired her to try it in her trendy downtown restaurant. She used some of the freshly sprouted beans in one of her daily specials.

Was the food prepped correctly? _____

Why or why not? _____

_____

⑤ The Dock House is known for its shrimp and hot wings. To make these items, fry cooks dip both the raw shrimp and the raw chicken wings in the same container of their secret batter. Then they deep-fry both food items until they are golden brown.

Was the food prepped correctly? _____

Why or why not? _____

_____

⑥ Norris wanted to make a special treat for the residents at the nursing home where he works. He used unpasteurized shell eggs to make banana muffins.

Was the food prepped correctly? _____

Why or why not? _____

_____

Pick the Right Way to Prep Food

Place a ✔ next to the correct answer in each pair.

① To thaw frozen food:

_____ A Place the item on a prep table at room temperature.

_____ B Place the item in a cooler at 41°F (5°C) or lower.

② To preserve food by smoking it:

_____ A Make sure the item has been thawed before smoking it.

_____ B Make sure you have a variance from your local regulatory authority first.

③ When using TCS leftovers to make salads:

_____ A Make sure they were held at 41°F (5°C) or lower for 7 days or less.

_____ B Make sure they were held at 41°F (5°C) or lower for 10 days or less.

④ When using pooled eggs:

_____ A Cook them right after mixing or store them at room temperature.

_____ B Cook them right after mixing or store them at 41°F (5°C) or lower.

⑤ When packaging fresh fruit juice on-site for sale later:

_____ A Treat the juice according to an approved HACCP plan.

_____ B Treat the juice by freezing and thawing it before packaging.

⑥ When using ice as an ingredient:

_____ A You can use ice that was used to cool food.

_____ B You cannot use ice that was used to cool food.

For answers, please turn to page 7.22.

Cooking Food

The only way to reduce pathogens in food to safe levels is to cook it to its minimum internal temperature. This temperature is different for each food. Once reached, you must hold the food at this temperature for a specific amount of time. If a customer requests a lower temperature, you need to inform them of the potential risk of foodborne illness. Also be aware of special menu restrictions if you serve high-risk populations.

While cooking reduces pathogens in food, it does not destroy spores or toxins they may have produced. You still must handle food correctly before you cook it.

How to Check Temperatures

To make sure the food you are cooking has reached the right temperature, you must know how to take the temperature correctly. Follow these guidelines.

Pick a thermometer with a probe that is the right size for the food.

Check the temperature in the thickest part of the food.

Take at least two readings in different locations.

Cooking Requirements for Specific Food

Minimum temperatures have been developed for TCS food. These temperatures are listed below. However, your operation or area might require different temperatures.

Cooking Requirements for Specific Types of Food	
Minimum Internal Temperature	Type of Food
165°F (74°C) for 15 seconds	• Poultry—including whole or ground chicken, turkey, or duck • Stuffing made with TCS ingredients • Stuffed meat, seafood, poultry, or pasta • Dishes that include previously cooked, TCS ingredients (Raw ingredients should be cooked to their minimum internal temperatures.)
155°F (68°C) for 15 seconds	• Ground meat—including beef, pork, and other meat • Injected meat—including brined ham and flavor-injected roasts • Mechanically tenderized meat • Ground seafood—including chopped or minced seafood • Eggs that will be hot-held for service
145°F (63°C) for 15 seconds	• Seafood—including fish, shellfish, and crustaceans • Steaks/chops of pork, beef, veal, and lamb • Eggs that will be served immediately
145°F (63°C) for 4 minutes	• Roasts of pork, beef, veal, and lamb
135°F (57°C)	• Commercially processed, ready-to-eat-food that will be hot-held for service (cheese sticks, deep-fried vegetables)
135°F (57°C)	• Fruit, vegetables, grains (rice, pasta), and legumes (beans, refried beans) that will be hot-held for service

Check your local regulatory requirements.

Cooking TCS Food in the Microwave

Meat, seafood, poultry, and eggs that you cook in a microwave must be cooked to 165°F (74°C). In addition, you should follow these guidelines.

- Cover the food to prevent its surface from drying out.

- Rotate or stir it halfway through the cooking process, as shown in the photo at left, so the heat reaches the food more evenly.

- Let the covered food stand for at least two minutes after cooking to let the food temperature even out.

- Check the temperature in at least two places to make sure that the food is cooked through.

Partial Cooking During Preparation

Some operations partially cook food during prep and then finish cooking it just before service. You must follow the steps below if you plan to partially cook meat, seafood, poultry, or eggs or dishes containing these items.

❶ Do not cook the food for longer than 60 minutes during initial cooking.

❷ Cool the food immediately after initial cooking.

❸ Freeze or refrigerate the food after cooling it. If refrigerating the food, make sure it is held at 41°F (5°C) or lower.

❹ Heat the food to at least 165°F (74°C) before selling or serving it.

❺ Cool the food if it will not be served immediately or held for service.

Your local regulatory authority may require you to have written procedures that explain how the food cooked by this process will be prepped and stored. These procedures must be approved by the regulatory authority and describe the following.

- How the requirements will be monitored and documented

- Which corrective actions will be taken if requirements are not met

- How these food items will be marked after initial cooking to indicate that they need further cooking

- How these food items will be separated from ready-to-eat food during storage, once initial cooking is complete

Consumer Advisories

You must cook TCS food to the minimum internal temperatures listed in this chapter—unless a customer requests otherwise. This might happen often in your operation, particularly if you serve meat, eggs, or seafood.

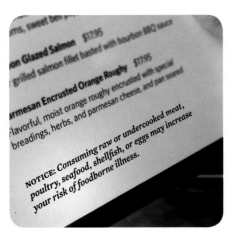

Disclosure   If your menu includes TCS items that are raw or undercooked, you must note it on the menu next to these items.

Reminder   You must advise customers who order food that is raw or undercooked of the increased risk of foodborne illness. You can do this by posting a notice in your menu, as shown in the photo at left. You can also provide this information using brochures, table tents, or signs. **Check your local regulatory requirements.**

Children's Menus

The FDA advises against offering raw or undercooked meat, poultry, seafood, or eggs to children. This is especially true for undercooked ground beef, which may be contaminated with shiga toxin-producing *E. coli* O157:H7.

Operations That Mainly Serve High-Risk Populations

Operations that mainly serve a high-risk population, such as nursing homes or day-care centers, cannot serve certain items. NEVER serve raw seed sprouts or raw or undercooked eggs, meat, or seafood. Examples include over-easy eggs, raw oysters on the half shell, and rare hamburgers.

Apply Your Knowledge

How Do You Check It?

Choose the picture that shows the right way to check a temperature. Write your answer in the space provided.

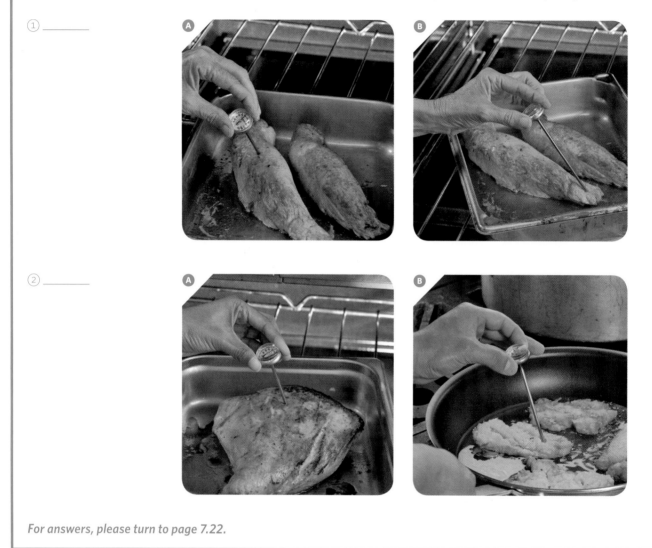

① _____

② _____

For answers, please turn to page 7.22.

What's the Temperature?

Identify the minimum internal cooking temperature for each food. Write the letter in the space provided. Some letters will be used more than once.

Ⓐ 135°F (57°C)

Ⓑ 145°F (63°C)

Ⓒ 155°F (68°C)

Ⓓ 165°F (74°C)

① _____ Salmon steak

② _____ Roasted vegetables that will be hot-held

③ _____ Ground pork

④ _____ Lamb chops

⑤ _____ Eggs for immediate service

⑥ _____ Duck

⑦ _____ Precooked, frozen cheese sticks

⑧ _____ Beef steak

⑨ _____ Chicken enchiladas made with previously cooked chicken

⑩ _____ Pork loin injected with marinade

⑪ _____ Broccoli cooked in a microwave

Consumer Advisory

Place a ✔ next to the correct answer in each pair.

① If a menu includes TCS items that are raw or undercooked:

_____ A You do not need to note it on the menu if you mainly serve high-risk populations.

_____ B You must note it on the menu next to each raw or undercooked item.

② If an operation mainly serves high-risk populations:

_____ A It cannot serve raw or undercooked eggs, meat, or seafood.

_____ B It can only serve raw or undercooked food if it posts a consumer advisory.

For answers, please turn to page 7.22.

Cooling and Reheating Food

When you don't serve cooked food immediately, you must get it out of the temperature danger zone as quickly as possible. That means cooling it quickly. You also need to reheat it correctly especially if you are going to hold it.

Cooling Food

As you know, pathogens grow well in the temperature danger zone. But they grow much faster at temperatures between 125°F and 70°F (52°C and 21°C). Food must pass through this temperature range quickly to reduce this growth.

Cool TCS food from 135°F (57°C) to 41°F (5°C) or lower within six hours.

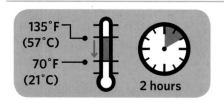

First, cool food from 135°F to 70°F (57°C to 21°C) within two hours.

Then cool it to 41°F (5°C) or lower in the next four hours.

If food has not reached 70°F (21°C) within two hours, it must be thrown out or reheated and then cooled again.

If you can cool the food from 135°F to 70°F (57°C to 21°C) in less than two hours, you can use the remaining time to cool it to 41°F (5°C) or lower. However, the total cooling time cannot be longer than six hours. For example, if you cool food from 135°F to 70°F (57°C to 21°C) in one hour, you have the remaining five hours to get the food to 41°F (5°C) or lower. ***Check your local regulatory requirements.***

How This Relates to Me

What are the time and temperature requirements for cooling food in your area?

_____

_____

_____

Methods for Cooling Food

The following factors affect how quickly food will cool.

Thickness or density of the food   The denser the food, the more slowly it will cool. For example, refried beans will take longer to cool than vegetable broth because the beans are denser than the broth.

Storage container   Stainless steel transfers heat away from food faster than plastic. Shallow pans let the heat from food disperse faster than deep pans.

NEVER place large quantities of hot food in a cooler to cool. Coolers are designed to keep cold food cold. Most are not designed to cool hot food quickly. Also, placing hot food in a cooler or freezer to cool it may not move the food through the temperature danger zone quickly enough.

Before cooling food, you should start by reducing its size. This will let it cool faster. Cut large food items into smaller pieces. Divide large containers of food into smaller containers or shallow pans. The foodhandler in the photo at left is doing this.

Ice-water bath   After dividing food into smaller containers, place them in a clean prep sink or large pot filled with ice water.

Stir the food frequently to cool it faster and more evenly.

Ice paddle   Plastic paddles are available that can be filled with ice or with water and then frozen. Food stirred with these paddles will cool quickly.

Food cools even faster when placed in an ice-water bath and stirred with an ice paddle.

Blast chiller or a tumble chiller
Blast chillers blast cold air across food at high speeds to remove heat. They are typically used to cool large amounts of food.

Tumble chillers tumble bags of hot food in cold water. Tumble chillers work well on thick food such as mashed potatoes.

Food can also be cooled by adding ice or cold water as an ingredient. This works for soups, stews, and other recipes that have water as an ingredient. When cooling this way, the recipe is made with less water than required. Cold water or ice is then added after cooking to cool the food and provide the remaining water.

Something to Think About...

That's Cool!

A small restaurant chain noted the same problem happening over and over at one of its operations. The restaurant was having trouble cooling its chili. The staff was packing it in five-gallon buckets before storing it in the cooler. Even though they filled the buckets only half full and used an ice paddle, the chili did not cool fast enough.

Management and the head chef set to work to find a solution. They decided that the best approach was to pour the chili into shallow pans before placing it in the cooler. The new procedure cooled the chili to 41°F (5°C) in a little less than two hours.

Once everyone tried this, the new procedure even earned praise from staff. They had been using the hotel pans to reheat the chili anyway. The new approach saved scraping and washing the buckets. The pans were also much easier to lift.

Reheating Food

How you reheat food depends on how you intend to use the food. Follow these guidelines.

Food reheated for hot-holding   From start to finish, you must heat the food to an internal temperature of 165°F (74°C) within two hours. Make sure the food stays at this temperature for at least 15 seconds. The foodhandler in the photo at left is reheating clam chowder for hot-holding. Roasts can be reheated to the alternative temperatures listed below, depending on the type of roast and the oven used. ***Check your local regulatory requirements.***

Temperature	Time
130°F (54°C)	112 minutes
131°F (55°C)	89 minutes
133°F (56°C)	56 minutes
135°F (57°C)	36 minutes
136°F (58°C)	28 minutes
138°F (59°C)	18 minutes
140°F (60°C)	12 minutes
142°F (61°C)	8 minutes
144°F (62°C)	5 minutes
145°F (63°C)	4 minutes

Food reheated for immediate service   You can reheat food that will be served immediately, like beef for a beef sandwich, to any temperature. However, you must make sure the food was cooked and cooled correctly.

Apply Your Knowledge

Is It Cool Enough?

Decide if the food in each situation is safe to serve. Explain why or why not.

① When the lunch buffet ended at 2:00 p.m., Marta started breaking it down. First, she removed the pan of vegetable curry that was held at 135°F (57°C) on the steam table. She put it on a cart while she got the other pans from the buffet. At 2:15 p.m., Marta divided the leftover vegetable curry into two bags and sealed them. Then she placed them into an ice-water bath. At 3:45 p.m., she checked their temperature. The vegetable curry was 65°F (18°C). Marta labeled the bags with the contents, temperature, and time, and put them in the cooler. At 5:30 p.m., when the dinner chef came in, he checked the temperature of the bags. They were 40°F (4°C).

Is the vegetable curry safe to serve? _____

Why or why not? _____

_____

_____

② Bill placed a stockpot of soup that had been held at 135°F (57°C) in an ice-water bath to cool at 1:00 p.m. At 3:00 p.m., he checked the temperature and found that it was 90°F (32°C). Bill added more ice to the ice-water bath, stirring it occasionally. At 4:00 p.m., when the soup had reached 70°F (21°C), he poured it into shallow pans and placed it in the cooler.

Is the soup safe to serve? _____

Why or why not? _____

_____

_____

Is It Hot Enough?

Decide if the food in each situation is safe to serve. Explain why or why not.

① At 9:00 a.m., Lin clocked in, said hello to her manager, and started to set up the buffet. Fifteen minutes later, she headed to the walk-in cooler, where she grabbed a stockpot of chili that had been made a few days earlier. She placed the stockpot on the stove and started reheating it. At 11:30 a.m., she checked the temperature of the chili, which had reached 155°F (68°C). Satisfied, she moved on to her next task.

Is the chili safe to serve? _____

Why or why not? _____

_____

_____

② Thursday at lunch, a customer ordered a French dip sandwich. Mina took a piece of roast from the cooler and sliced the beef. She put the slices in a hot pan of au jus and heated it for a few minutes while she returned the roast to the cooler. Then she made the sandwich, plated it, and placed it on the counter for pickup.

Is the sandwich safe to serve? _____

Why or why not? _____

_____

_____

For answers, please turn to page 7.23.

Chapter Summary

To protect food during preparation, you must handle it safely. The keys are time and temperature control and preventing cross-contamination.

Thaw frozen food in the cooler, under running water, in a microwave oven, or as part of the cooking process. Never thaw food at room temperature. Employees should prep food in small batches. They must also keep their workstations and utensils clean and sanitized. Prepped food that isn't going to be cooked immediately should be put back in the cooler.

Cooking can reduce pathogens in food to safe levels. You must cook food to minimum internal temperatures for a specific amount of time. These temperatures vary from product to product. Cooking does not kill the spores or toxins some pathogens produce. That is why it is so important to handle food correctly.

Once food is cooked, it should be served as quickly as possible. If it is going to be stored and served later, it must be cooled rapidly. TCS food must be cooled from 135°F to 70°F (57°C to 21°C) within two hours. Then it must be cooled from 70°F to 41°F (21°C to 5°C) or lower in the next four hours.

Before food is cooled, it should be reduced in size. Cut large food items into smaller pieces. Divide large containers of food into smaller ones. There are several ways to cool food safely. You can use an ice-water bath, stir food with ice paddles, or use a blast or tumble chiller.

Reheated TCS food that will be hot-held must be heated to an internal temperature of 165°F (74°C) within two hours. Make sure the food stays at this temperature for 15 seconds.

Chapter Review Case Study

Now take what you have learned in this chapter and apply it to the following case study.

When Angie arrived for work at Sunnydale Nursing Home, she knew she had a busy day ahead of her. Looking in the freezer, Angie realized that she had forgotten to thaw the chicken breasts she planned to serve for dinner. Moving quickly, she placed the frozen chicken in a prep sink and turned on the hot water.

Once the chicken breasts had thawed, making dinner was a snap. By 7:30 p.m., all the residents had eaten dinner. As Angie began cleaning up, she realized she had a lot of cooked chicken breasts left over. Betty, the new assistant manager, had forgotten to tell Angie that several residents were going to a local festival and would miss dinner. "No problem," Angie thought. "We can use the leftover chicken to make chicken salad."

Angie left the still-hot chicken breasts in a pan on the prep table while she started putting other food away and cleaning up. At 9:45 p.m., when everything else was clean, she put her hand over the pan of chicken breasts and decided they were cool enough to handle. She covered the pan with plastic wrap and put it in the cooler.

Three days later, Angie came in to work on the early shift. She decided to make chicken salad from the leftover chicken breasts. Angie took all the ingredients she needed for chicken salad out of the cooler and put them on a prep table. Then she started breakfast.

First, she cracked three-dozen eggs into a large bowl, added some milk, and set the bowl near the stove. Then she took bacon out of the cooler and put it on the prep table next to the chicken-salad ingredients. She peeled off strips of bacon onto a sheet pan and put the pan into the oven. Then she went back to the stove to whisk the eggs and pour them onto the griddle. When they looked ready, Angie checked the temperature. The eggs had reached 145°F (63°C). Angie scooped the scrambled eggs into a hotel pan and put it in the steam table.

As soon as breakfast was cooked, Angie went back to the prep table to wash and chop celery and cut up the chicken for the salad.

① What did Angie do wrong?

_____

_____

_____

_____

② What should Angie have done differently?

_____

_____

_____

_____

For answers, please turn to page 7.23.

Study Questions

Circle the best answer to each question below.

① **Beef stew must be cooled from 135°F to 70°F (57°C to 21°C) within _____ hours and from 70°F to 41°F (21°C to 5°C) or lower in the next _____ hours.**

A 2, 4

B 3, 2

C 4, 2

D 2, 3

② **What must you do immediately after thawing food in a microwave?**

A Hold it.

B Cook it.

C Cool it.

D Freeze it.

③ **What is the minimum internal cooking temperature for stuffed pork chops?**

A 135°F (57°C)

B 145°F (63°C)

C 155°F (68°C)

D 165°F (74°C)

④ **What are the time and temperature requirements for reheating TCS food for hot-holding?**

A 135°F (57°C) for 15 seconds within 2 hours

B 145°F (63°C) for 15 seconds within 2 hours

C 155°F (68°C) for 15 seconds within 2 hours

D 165°F (74°C) for 15 seconds within 2 hours

⑤ **What is the minimum internal cooking temperature for eggs, meat, poultry, and seafood cooked in a microwave?**

A 135°F (57°C)

B 145°F (63°C)

C 155°F (68°C)

D 165°F (74°C)

⑥ **What is the minimum internal cooking temperature for eggs that will be hot-held for later service?**

A 135°F (57°C)

B 145°F (63°C)

C 155°F (68°C)

D 165°F (74°C)

⑦ **What is the danger of NOT cleaning and sanitizing a prep table between uses?**

A Off flavors in food

B Poor personal hygiene

C Cross-contamination

D Time-temperature abuse

⑧ **What is the correct way to cool a stockpot of clam chowder?**

A Put the stockpot into a cooler.

B Put the stockpot into a freezer.

C Put the stockpot into ice water.

D Put the stockpot on a prep table.

⑨ **What is the danger when thawing food at room temperature?**

A Cross-contamination

B Poor personal hygiene

C Physical contamination

D Time-temperature abuse

⑩ **What is the minimum internal cooking temperature for ground beef?**

A 135°F (57°C)

B 145°F (63°C)

C 155°F (68°C)

D 165°F (74°C)

For answers, please turn to page 7.24.

Answers

7.6   **What's the Problem?**

① No. One case of green onions could cross-contaminate the other. Between batches, he should have emptied the sink, cleaned and sanitized it, and changed the ice water.

② No. The meat and cheese are being time-temperature abused. She should only take out of the cooler what she can use within a short amount of time.

③ Yes. He used separate equipment for the meat and the produce.

④ No. Jessica should have contacted her local regulatory authority before sprouting the beans. She may have needed a variance to do this.

⑤ No. You should never use the same batter for two types of food if one of them can cause an allergic reaction.

⑥ Yes. Although the muffins were made for a high-risk population, they were cooked thoroughly. Therefore, he did not need to use pasteurized eggs.

7.7   **Pick the Right Way to Prep Food**

① B   ④ B

② B   ⑤ A

③ A   ⑥ B

7.12   **How Do You Check It?**

① A

② A

7.13   **What's the Temperature?**

① B   ⑦ A

② A   ⑧ B

③ C   ⑨ D

④ B   ⑩ C

⑤ B   ⑪ A

⑥ D

7.13   **Consumer Advisory**

① B

② A

7.17   Is It Cool Enough?

① Yes. The vegetable curry was cooled to 70°F (21°C) or lower within the first two hours. Then it was cooled to 41°F (5°C) or lower within the next four hours.

② No. The soup was not cooled to 70°F (21°C) within the first two hours.

7.17   Is It Hot Enough?

① No. The chili did not reach an internal temperature of 165°F (74°C) within two hours.

② Yes. Assuming the roast beef was cooked and cooled correctly, it can be reheated to any temperature since it is being served immediately.

7.18   Chapter Review Case Study

① Here is what Angie did wrong.

- She thawed the chicken breasts the wrong way. She should not have thawed them under hot water.

- She cooled the leftover chicken breasts the wrong way. She should not have left them out to cool at room temperature.

- She subjected the chicken salad ingredients to time-temperature abuse when she left them on the prep table.

- She pooled shell eggs when prepping the scrambled eggs. If shell eggs are going to be pooled when serving a high-risk population, such as the residents of a nursing home, they must be pasteurized. This became a bigger problem when Angie undercooked the eggs.

- She did not handle the pooled eggs right. She left the bowl in the temperature danger zone by putting it near a warm stove.

- She did not cook the scrambled eggs to the right temperature before storing them on the steam table.

- She did not clean and sanitize the prep table after she made the bacon and before she chopped the celery and chicken.

Continued on the next page ▶

► *Continued from previous page*

② Here is what Angie should have done differently.

- If Angie needed to thaw the chicken breasts quickly, she should have either used a microwave or placed them under running water at 70°F (21°C) or lower.

- To cool the chicken breasts quickly for two-stage cooling, she could have used a blast chiller or placed the container of chicken breasts in an ice-water bath. Then she could move them to the cooler.

- She should have left the chicken salad ingredients in the cooler until she was ready to prep the salad.

- She should have used pasteurized shell eggs or egg products.

- She should have held the eggs in the cooler until she was ready to cook them.

- She should have cooked the eggs to be held for later service to at least 155°F (68°C) for 15 seconds.

- She should have cleaned and sanitized the prep table after she made the bacon and before she chopped the celery and chicken.

7.20   Study Questions

① A		⑥ C	
② B		⑦ C	
③ D		⑧ C	
④ D		⑨ D	
⑤ D		⑩ C	

Notes

_____

_____

_____

_____

_____

_____

_____

_____

_____

_____

_____

_____

_____

_____

_____

_____

_____

_____

_____

_____

_____

_____

_____

_____

_____

_____

_____

_____

_____

_____

_____

_____

_____

8

The Flow of Food: Service

In the News

Country Club Sickens 67

A self-service buffet served to 855 people at a country club made 67 people sick. *Salmonella* spp. caused the outbreak. Health officials traced the pathogen to turkey, stuffing, and gravy served at the buffet.

Officials learned that the turkey had been cooked the right way but was held for five hours at room temperature after cooking, which allowed the pathogen to grow. They also believed that customers used the same utensils for the turkey and other food during service.

You Can Prevent This

The operation in the story above had two problems in its self-service area. It failed to hold the turkey at the right temperature to prevent pathogen growth. It also let customers use the same utensils for more than one item, causing cross-contamination.

Holding and serving food the right way would have helped prevent this outbreak. In this chapter, you will learn guidelines for keeping food safe after you have prepped and cooked it. These guidelines include the following.

- Holding hot food
- Holding cold food
- Using time as a method of control for food
- Preventing contamination of food in self-service areas and when serving food to customers

---

Concepts from Earlier Chapters

Before reading this chapter, remember these concepts and facts.

Temperature danger zone   Temperature range between 41°F and 135°F (5°C to 57°C). Foodborne pathogens grow well in it.

Cross-contamination   Transfer of pathogens from one surface or food to another.

TCS food   Food that requires time and temperature control for safety.

Holding Food

Food that is being held for service is at risk for time-temperature abuse and cross-contamination. If your operation holds food, you must make policies that reduce these risks. Focus on time and temperature control, but don't forget about protecting the food from contamination. In some cases, you might even be able to hold food without controlling its temperature.

Guidelines for Holding Food

Follow these guidelines when holding food.

PATHOGEN PREVENTION ☠

Temperature   Hold TCS food at the right internal temperature.

- Hold hot food at 135°F (57°C) or higher. This will prevent pathogens such as *Bacillus cereus* from growing to unsafe levels.

- Hold cold food at 41°F (5°C) or lower. This will prevent pathogens such as *Staphylococcus aureus* from growing to unsafe levels.

Thermometer   Use a thermometer to check a food's internal temperature, as the foodhandler in the photo at left is doing. NEVER use the temperature gauge on a holding unit to do it. The gauge does not check the internal temperature of the food.

Time   Check food temperature at least every four hours.

- Throw out food that is not 41°F (5°C) or lower or 135°F (57°C) or higher.

- You can also check the temperature every two hours. This will leave time for corrective action.

Hot-holding equipment   NEVER use hot-holding equipment to reheat food unless it is built to do so. Most hot-holding equipment does not pass food through the temperature danger zone quickly enough. Reheat food correctly. Then move it to the holding unit.

Policies   Create policies about how long the operation will hold food. Also, create policies about when to throw away held food. For example, your policy may let you refill a pan of veal in a buffet all day, as long as you throw it out at the end of the day.

Food covers and sneeze guards   Cover food and install sneeze guards to protect food from contaminants. Covers, like the ones shown in the photo at left, also help maintain a food's internal temperature.

Check your local regulatory requirements.

Holding Food Without Temperature Control

Your operation may want to display or hold TCS food without temperature control. Here are some examples of when you might hold food without temperature control.

- When displaying food for a short time, such as at an off-site catered event, as shown in the photo at left

- When electricity is not available to power holding equipment

If your operation displays or holds TCS food without temperature control, it must do so under certain conditions. Also note that the conditions for holding cold food are different from those for holding hot food.

Cold Food

You can hold cold food without temperature control for up to six hours if you meet these conditions.

- Hold the food at 41°F (5°C) or lower before removing it from refrigeration.

- Label the food with the time you removed it from refrigeration and the time you must throw it out. The discard time on the label must be six hours from the time you removed the food from refrigeration, as shown in the photo at left. For example, if you remove potato salad from refrigeration at 3:00 p.m. to serve at a picnic, the discard time on the label should be 9:00 p.m. This equals six hours from the time you removed it from refrigeration.

- Make sure the food does not exceed 70°F (21°C) while it is being served. Throw out any food that exceeds this temperature.

- Sell, serve, or throw out the food within six hours.

Hot Food

You can hold hot food without temperature control for up to four hours if you meet these conditions.

- Hold the food at 135°F (57°C) or higher before removing it from temperature control.

- Label the food with the time you must throw it out. The discard time on the label must be four hours from the time you removed the food from temperature control, as shown in the photo at left.

- Sell, serve, or throw out the food within four hours.

Before using time as a method of control, check with your local regulatory authority for specific requirements.

How This Relates to Me

Does your area allow you to hold ready-to-eat TCS food without temperature control?

_____

If yes, what are the requirements for doing so?

_____

_____

_____

Apply Your Knowledge

To Serve or Not to Serve?

Place a ✔ next to the food items that were handled without temperature control correctly.

① _____ Potato salad that will be held without temperature control is taken out of refrigeration at 10:00 a.m. and labeled: Removed at 10:00 a.m.; throw out at 4:00 p.m.

② _____ Cold, sliced deli meat that was held at 50°F (10°C) the night before is served at room temperature for 6 hours and then thrown out at the time shown on the label.

③ _____ A pan of chicken salad held at room temperature for 5 hours is thrown out after its temperature was checked and found to be 70°F (21°C).

④ _____ Fried chicken held at 135°F (57°C) is placed in chafing dishes and labeled at 11:00 a.m. It is served to nursing-home residents on a hot-holding unit for 4 hours and then thrown out.

⑤ _____ Broccoli casserole held at 140°F (60°C) is labeled and placed on display at noon. It is served at room temperature and then thrown out at 2:30 p.m.

For answers, please turn to page 8.12.

Serving Food

The biggest threat to food that is ready to be served is contamination. Your kitchen and service staff must know how to serve food in ways that keep it safe. Dining rooms, self-service areas, off-site locations, and vending machines all have specific guidelines that staff must follow.

Kitchen Staff Guidelines

Train your kitchen staff to serve food in these ways.

Bare-hand contact with food Handle ready-to-eat food with tongs, deli sheets, or gloves. The photo at left shows two ways to minimize bare-hand contact.

Some regulatory authorities allow bare-hand contact if the operation has received prior approval. The operation must outline policies for the following areas.

- Employee health

- Employee training in handwashing and personal hygiene

Check your local regulatory requirements.

How This Relates to Me

Does your regulatory authority allow bare-hand contact with ready-to-eat food?

_____

If yes, what are the requirements for doing so?

_____

_____

Clean and sanitized utensils Use separate utensils for each food item. Clean and sanitize them after each serving task. If using utensils continuously, clean and sanitize them at least once every four hours.

Serving utensils Store serving utensils in the food with the handle extended above the rim of the container, as shown in the photo at left. You can also place them on a clean, sanitized food-contact surface. Spoons or scoops used to serve food such as ice cream or mashed potatoes can be stored under running water that is 135°F (57°C).

Service Staff Guidelines

Service staff must be as careful as kitchen staff. They can contaminate food simply by handling the food-contact areas of glasses, dishes, and utensils. Service staff should use these guidelines when serving food.

Hold dishes by the bottom or edge.

Hold glasses by the middle, bottom, or stem.

Do **NOT** touch the food-contact areas of dishes or glassware.

Carry glasses in a rack or on a tray to avoid touching the food-contact surfaces.

Stacking china and glassware can cause them to chip and break.

Do **NOT** stack glasses when carrying them.

Hold flatware by the handle.

Store flatware so servers grasp handles, not food-contact surfaces.

Do **NOT** hold flatware by food-contact surfaces.

Minimize bare-hand contact with food that is ready to eat.

Use ice scoops or tongs to get ice.

NEVER scoop ice with your bare hands or a glass. A glass may chip or break.

Preset Tableware

If your operation presets tableware on dining tables, you must take steps to prevent it from becoming contaminated. This might include wrapping or covering the items.

Table settings do not need to be wrapped or covered if extra settings meet these requirements.

- They are removed when guests are seated.
- If they remain on the table, they are cleaned and sanitized after guests have left.

Re-serving Food

Service and kitchen staff should also know the rules about re-serving food previously served to a customer.

Menu items   Do **NOT** re-serve food returned by one customer to another customer.

Condiments   **NEVER** re-serve uncovered condiments. Do **NOT** combine leftover condiments with fresh ones, like the foodhandler in the photo at left is doing. Throw away opened portions or dishes of condiments after serving them to customers. Salsa, butter, mayonnaise, and ketchup are examples.

Bread or rolls   Do **NOT** re-serve uneaten bread to other customers. Change linens used in bread baskets after each customer.

Garnishes   **NEVER** re-serve plate garnishes, such as fruit or pickles, to another customer. Throw out served but unused garnishes.

Prepackaged food   In general, you may re-serve only unopened, prepackaged food. These include condiment packets, wrapped crackers, or wrapped breadsticks. You may re-serve bottles of ketchup, mustard, and other condiments.

Self-Service Areas

Self-service areas can be contaminated easily. Follow these guidelines to prevent contamination and time-temperature abuse.

Sneeze guards   Protect food on display with sneeze guards or food shields. They should be located 14 inches (36 centimeters) above the counter and should extend 7 inches (18 centimeters) beyond the food, as shown in the photo at left.

Labels   Label containers located on self-service areas. Place the name of the food, such as types of salad dressing, on ladle handles.

Temperature   Keep hot food hot, 135°F (57°C) or higher. Keep cold food cold, 41°F (5°C) or lower.

Raw and ready-to-eat food   Keep raw meat, seafood, and poultry separate from ready-to-eat food in self-service areas.

PATHOGEN PREVENTION

Refills   Do not let customers refill dirty plates or use dirty utensils at self-service areas. Pathogens such as Norovirus can easily be transferred by reused plates and utensils. Assign a staff member to hand out clean plates for return visits. Post signs with polite tips about self-service etiquette.

Ice   Ice used to keep food or beverages cold should NEVER be used as an ingredient.

Off-Site Service

The longer the delay from preparation to the point of service, the greater the risk that food will be exposed to contamination or time-temperature abuse. To transport food safely, follow these procedures.

Food containers   Pack food in insulated food containers that can keep food at 135°F (57°C) or higher, or at 41°F (5°C) or lower. Use only food-grade containers, such as those shown in the photo at left. They should be designed so food cannot mix, leak, or spill.

At the service site, use appropriate containers or equipment to hold food at the right temperature.

Delivery vehicles   Clean the inside of delivery vehicles regularly.

Internal temperature   Check internal food temperatures. If containers or delivery vehicles are not holding food at the right temperature, reevaluate the length of the delivery route or the efficiency of the equipment being used.

Labels   Label food with a use-by date and time, and reheating and service instructions for staff at off-site locations. This is shown in the photo at left.

Utilities   Make sure the service site has the right utilities.

- Safe water for cooking, dishwashing, and handwashing

- Garbage containers stored away from food-prep, storage, and serving areas

Storage   Store raw meat, seafood, and poultry and ready-to-eat items separately. For example, store raw chicken separately from ready-to-eat salads.

Vending Machines

Handle food prepped and packaged for vending machines with the same care as any other food served to customers. Vending operators should protect food from contamination and time-temperature abuse during transport, delivery, and service. To keep vended food safe, follow these guidelines.

- Check product shelf life daily. Products often have a code date, such as an expiration or use-by date, like that shown in the photo at left. If the date has expired, throw out the food immediately. Throw out refrigerated food prepped on-site if not sold within seven days of preparation.

- Keep TCS food at the right temperature. It should be held at 41°F (5°C) or lower, or at 135°F (57°C) or higher.

- Dispense TCS food in its original container.

- Wash and wrap fresh fruit with edible peels before putting it in a machine.

Apply Your Knowledge

Re-Serve or Throw Out?

Write a **T** next to the food that you must throw out. Write an **R** next to the items you can re-serve.

① _____ Labeled chili held without temperature control for 5 hours

② _____ Previously served, but untouched, basket of bread

③ _____ Bottle of ketchup

④ _____ Untouched slice of pie with whipped cream returned by a customer

⑤ _____ Individually wrapped crackers

⑥ _____ Unwrapped butter served on a plate

⑦ _____ Mustard packets

⑧ _____ Ice used to hold cold food on a self-service area

⑨ _____ Breaded, baked fish returned by a customer who wanted broiled fish

⑩ _____ Slice of apple pie that has been in a vending machine for 6 days

For answers, please turn to page 8.12.

Chapter Summary

Safe foodhandling does not stop once food is prepped and cooked. You must continue to protect it from time-temperature abuse and contamination until it is served. When holding TCS food for service, keep hot food at 135°F (57°C) or higher. Keep cold food at 41°F (5°C) or lower. Check the internal temperature of food at least every four hours. Throw it out if it is not at the right temperature. Train employees to avoid cross-contamination when handling service items and tableware. Teach them about the potential hazards from re-serving food such as plate garnishes, breads, or open dishes of condiments.

Self-service areas can be easily contaminated. Post signs about self-service rules. Assign employees to hand out clean plates for refills. Protect food on display with sneeze guards. Make sure equipment holds food at the right temperature.

Follow safety procedures when prepping, delivering, or serving food off-site. Containers and delivery vehicles must hold food at the right temperature. Make sure the service site has the right utilities. Additionally, handle vending machine food as carefully as any other food. Check product shelf-life daily. Hold TCS food at the right temperature. Wash and wrap fresh fruit before putting it in a machine.

Chapter Review Case Study

Now take what you have learned in this chapter and apply it to the following case study.

Jill, a line cook on the morning shift at Memorial Hospital, was busy helping the kitchen staff put food on display for lunch in the hospital cafeteria. Ann, the kitchen manager who usually supervised lunch in the cafeteria, was at an all-day seminar on food safety. Jill was also responsible for making sure meals were trayed and put into food carts for transport to the patients' rooms. The staff also packed two dozen meals each day for a neighborhood group that delivered them to homebound elderly people.

First, Jill looked for insulated food containers for the delivery meals. When she could not find them, she loaded the meals into cardboard boxes she found near the back door, knowing the driver would arrive soon to pick them up. To help the cafeteria staff, Jill filled a pan with soup by dipping a two-quart measuring cup into the stockpot and pouring the soup into the pan. She carried the pan out to the cafeteria, put it into the steam table, and turned the thermostat on low.

The lunch hour was hectic. The cafeteria was busy, and the staff had many meals to tray and deliver. Halfway through lunch, a cashier came back to the kitchen to tell Jill that the salad bar needed replenishing. Since she was busy, Jill asked a kitchen employee to take pans of prepped ingredients out of the refrigerator and put them on the salad bar. When she looked up a few moments later, she saw the kitchen employee send away two children who were eating carrot sticks from the salad bar.

With lunch almost over, Jill breathed a sigh of relief. She moved down the cafeteria serving line, checking food temperatures. One of the casseroles was at 130°F (54°C). Jill checked the water level in the steam table and turned up the thermostat, and then went to clean up the kitchen and finish her shift.

① What did Jill do wrong?

_____

_____

② What should Jill have done?

_____

_____

For answers, please turn to page 8.12.

Study Questions

Circle the best answer to each question below.

① **When serving, it is important to avoid touching the _____ of a plate.**

A top

B edges

C side

D bottom

② **Serving utensils should be used to serve a maximum of _____ food item(s) at a time.**

A 1

B 2

C 3

D 4

③ **At what maximum internal temperature should cold TCS food be held?**

A 0°F (–17°C)

B 32°F (0°C)

C 41°F (5°C)

D 60°F (16°C)

④ **When returning to self-service lines for more food, customers should not _____ their dirty plates.**

A carry

B overload

C refill

D stack

⑤ **At what minimum internal temperature should hot TCS food be held?**

A 115°F (46°C)

B 125°F (52°C)

C 135°F (57°C)

D 145°F (63°C)

⑥ **Where allowed, TCS hot food can be held without temperature control for a maximum of _____ hours before being sold, served, or thrown out.**

A 2

B 4

C 6

D 8

For answers, please turn to page 8.12.

Answers

8.4   To Serve or Not to Serve?

1, 3, 4, and 5 should be marked.

8.9   Re-Serve or Throw Out?

① T		⑥ T	
② T		⑦ R	
③ R		⑧ T	
④ T		⑨ T	
⑤ R		⑩ R	

8.10   Chapter Review Case Study

① Here is what Jill did wrong.

- She packed the deliveries in cardboard boxes instead of rigid, insulated carriers.

- She used the wrong utensil to fill the soup pan.

- She did not assign a staff member to monitor the self-service area.

- She failed to make sure that the internal temperature of the food on the steam table was checked at least every four hours. This would have alerted her to the fact that the steam table was not maintaining the proper temperature and that the casserole was in the temperature danger zone.

② Here is what Jill should have done.

- She should have kept the delivery meals in a hot-holding cabinet or left the food in a steam table until suitable containers were found or the driver arrived.

- She should have used a long-handled ladle, which would have kept her hands away from the soup, preventing possible contamination, as she ladled it out.

- She should have made sure that an employee was assigned to monitor the food bar to ensure that customers, such as the children, followed proper etiquette.

- She should have thrown out the casserole and any other food that was not at the right temperature, since she did not know how long the food was in the temperature danger zone.

8.11   Study Questions

① A		④ C	
② A		⑤ C	
③ C		⑥ B	

Notes

III Food Safety Management Systems, Facilities, and Pest Management

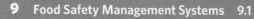

9

Food Safety Management Systems

In the News

Blue Skies Handles It Right

The calls started on a Thursday morning at Blue Skies Café, a small but well-liked diner in a busy city neighborhood. The callers complained of stomach cramps and diarrhea. The owner of the café, Linda Burke, took the first few calls and realized that she might have a foodborne-illness outbreak on her hands. She filled out an incident report for each call, and then she contacted the local health department.

"We were also getting calls, so we went to the café to see what happened," said José Perez, the health inspector assigned to the case. "With the cooperation of Ms. Burke, we were able to identify the Caesar dressing from the day before the outbreak as the source of the customers' illnesses."

A batch of the Caesar dressing was made that Wednesday with contaminated eggs, eventually making 30 people sick. Because Caesar dressing isn't fully cooked, the café could not have done anything different to prep the dressing. "To correct the issue, we now use pasteurized eggs for the dressing, and we make new batches every few hours," said Ms. Burke.

Mr. Perez also noted that the café's health-inspection score was not changed because of the outbreak, nor was the operation forced to close. "They handled the problem quickly, and the rest of the operation is clean and well run," he said. Additionally, the café's insurance covered the healthcare costs and lost wages that the outbreak caused.

You Can Prevent This

A foodborne-illness outbreak is any manager's nightmare, but as you can see in the story above, you can survive one. Creating a food safety management system will help prevent problems. A crisis-management plan will help you manage an outbreak if one happens.

In this chapter, you'll learn how to build and apply these systems.

- Food safety management system
- Crisis-management plan

Concepts from Earlier Chapters

Before reading this chapter, remember these concepts and facts.

Approved, reputable supplier   Supplier that has been inspected and complies with applicable local, state, and federal laws.

Food Safety Management Systems

In chapters 5 through 8, you learned how to handle food safely throughout the flow of food. Now, you will learn how all of this information can be applied to a food safety management system.

To do this, you must understand how a food safety management system works. The two primary systems are active managerial control and Hazard Analysis Critical Control Point (HACCP).

Overview of Food Safety Management Systems

A food safety management system is a group of procedures and practices intended to prevent foodborne illness. It does this by actively controlling risks and hazards throughout the flow of food. Active managerial control and HACCP are two ways to build a system.

Having some food safety programs already in place gives you the foundation for your system. The principles presented in the ServSafe program are the basis of these programs. Here are some examples of the programs your operation needs.

Personal hygiene program

Supplier selection and specification program

Sanitation and pest-control programs

Facility design and equipment-maintenance program

Food safety training program

Active Managerial Control

Active managerial control is one way to manage food safety risks in your operation. This approach focuses on controlling the five most common risk factors that cause foodborne illness, as identified by the Centers for Disease Control and Prevention (CDC).

❶ Purchasing food from unsafe sources

❷ Failing to cook food adequately

❸ Holding food at incorrect temperatures

❹ Using contaminated equipment

❺ Practicing poor personal hygiene

The *FDA Food Code* has identified five ways to control these risks. It has created the following public-health interventions to protect consumer health.

Demonstration of knowledge   As a manager, you must be able to show that you know what to do to keep food safe. One example is knowing the illnesses that foodborne pathogens can cause.

Staff health controls   Staff health controls are policies and procedures that you put into place to make sure your employees are practicing personal hygiene. One example is the exclusion and restriction criteria you learned in chapter 4.

Controlling hands as a vehicle of contamination   These controls help prevent cross-contamination from hands to food. Using tongs for ready-to-eat food is one example.

Time and temperature parameters for controlling pathogens   You must keep food out of the temperature danger zone. One example is following correct cooling procedures.

Consumer advisory   These are notices that you must provide to your customers about the risks of raw or undercooked food. One example is the notification you must place on your menu if it includes TCS items that are raw or undercooked.

Active Managerial Control Approach

To use active managerial control to manage food safety risks, you must follow these steps.

❶ Consider the five risk factors throughout the flow of food in your operation. Identify issues that could impact food safety. The Food Safety Self-Assessment in the appendix of this book may help you.

❷ Create policies and procedures that address the issues you identified. Consider asking your staff for suggestions as you create them. You also may need to provide training on these policies and procedures.

❸ Regularly monitor the policies and procedures that you developed, as shown in the photo at left. This step is critical to the system's success. It can help you determine if your policies and procedures are being followed. If not, you may have to revise them, create new ones, or retrain your staff.

❹ Verify that you are actually controlling the risk factors. Use feedback to adjust policies and procedures, so you can continuously improve the system. Feedback can come from both internal and external sources. Internal sources include records, temperature logs, and self-inspections. External sources include health-inspection reports, customer comments, and quality-assurance audits.

Example of Active Managerial Control

Here is an example of how one seafood restaurant chain used active managerial control.

❶ Considering the risk factors   The chain identified the purchasing of seafood from unsafe sources as a risk in their operations.

❷ Creating policies and procedures   The managers developed a list of approved seafood suppliers. They based the list on criteria that made sure the seafood they received would meet their safety and quality standards. Next, they created a policy that seafood could be purchased only from suppliers on this list.

❸ Monitoring the policies and procedures   The managers decided that all seafood invoices and deliveries would be monitored, as shown in the photo at left.

❹ Verifying the system   On a regular basis, the managers checked the standards they had created for choosing seafood suppliers. They wanted to make sure the standards were still able to control the risk. They also decided to review their policy regularly and change it as needed.

Something to Think About...

Get a Handle on It!

A local health department was inspecting a unit in a large quick-service chain. The inspector noticed that the grill operator handling raw chicken fillets also put cooked fillets in a holding drawer. A sandwich maker touched the handle of the drawer each time she retrieved a cooked fillet.

The health inspector saw that the grill operator was contaminating the holding drawer handle. It happened each time he put a cooked fillet inside—since his hands had touched raw chicken. When the sandwich maker touched the contaminated handle, there was a chance of cross-contamination.

Working with the unit manager, the inspector recommended adding an extra handle to the holding drawer. The grill operator and the sandwich maker were assigned their own handles. The chain adopted the recommendation in all of its units.

In dealing with the risk of contamination, the chain followed the procedure outlined by its active managerial control system. This included modifying their standard operating procedures (SOPs) to control the risk and retraining staff. They also incorporated the new SOPs in the chain's monitoring program.

HACCP

A HACCP system can also be used to control risks and hazards throughout the flow of food. HACCP (pronounced HASS-ip) is based on identifying significant biological, chemical, or physical hazards at specific points within a product's flow. Once identified, the hazards can be prevented, eliminated, or reduced to safe levels.

An effective HACCP system must be based on a written plan. This plan must be specific to each facility's menu, customers, equipment, processes, and operations. Since each HACCP plan is unique, a plan that works for one operation may not work for another.

The HACCP Approach

A HACCP plan is based on seven basic principles. They were created by the National Advisory Committee on Microbiological Criteria for Foods. These principles are the seven steps that outline how to create a HACCP plan.

The Seven HACCP Principles

Each HACCP principle builds on the information gained from the previous principle. You must consider all seven principles, in order, when developing your plan.

Here are the seven principles.

1. Conduct a hazard analysis.

2. Determine critical control points (CCPs).

3. Establish critical limits.

4. Establish monitoring procedures.

5. Identify corrective actions.

6. Verify that the system works.

7. Establish procedures for record keeping and documentation.

In general terms, the principles break into three groups.

- Principles 1 and 2 help you identify and evaluate your hazards.

- Principles 3, 4, and 5 help you establish ways for controlling those hazards.

- Principles 6 and 7 help you maintain the HACCP plan and system, and verify its effectiveness.

The next few pages provide an introduction to these principles. They also present an overview of how to build a HACCP program.

A real-world example has also been included for each principle. It shows the efforts of Enrico's, an Italian restaurant, as it implements a HACCP program. The example will appear immediately after the explanation of each principle.

Principle 1: Conduct a Hazard Analysis

First, identify and assess potential hazards in the food you serve. Start by looking at how food is processed in your operation. Many types of food are processed in similar ways. Here are some common processes.

- Prepping and serving without cooking (salads, cold sandwiches, etc.)

- Prepping and cooking for same-day service (grilled chicken sandwiches, hamburgers, etc.)

- Prepping, cooking, holding, cooling, reheating, and serving (chili, soup, pasta sauce with meat, etc.)

Look at your menu and identify items that are processed like this. Next, identify the TCS food. Determine where food safety hazards are likely to occur for each TCS food. There are many types of hazards to look for.

- Biological

- Chemical

- Physical

Principle 1 Example

The management team at Enrico's decided to implement a HACCP program. They began by analyzing their hazards.

The team noted that many of their dishes are received, stored, prepped, cooked, and served the same day. The most popular of these items was the spicy charbroiled chicken breast.

The team determined that bacteria were the most likely hazard to food prepped this way.

Principle 2: Determine Critical Control Points (CCPs)

Find the points in the process where the identified hazard(s) can be prevented, eliminated, or reduced to safe levels. These are the critical control points (CCPs). Depending on the process, there may be more than one CCP.

Principle 2 Example

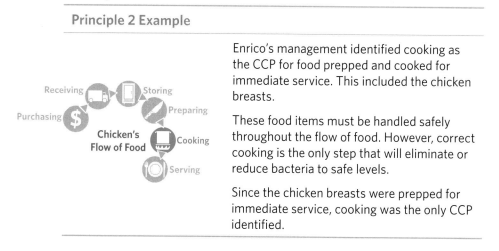

Enrico's management identified cooking as the CCP for food prepped and cooked for immediate service. This included the chicken breasts.

These food items must be handled safely throughout the flow of food. However, correct cooking is the only step that will eliminate or reduce bacteria to safe levels.

Since the chicken breasts were prepped for immediate service, cooking was the only CCP identified.

Principle 3: Establish Critical Limits

For each CCP, establish minimum or maximum limits. These limits must be met to prevent or eliminate the hazard, or to reduce it to a safe level.

Principle 3 Example

With cooking identified as the CCP for Enrico's chicken breasts, a critical limit was needed. Management determined that the critical limit would be cooking the chicken to a minimum internal temperature of 165°F (74°C) for 15 seconds.

They decided that the critical limit could be met by cooking chicken breasts in the broiler for 16 minutes.

Principle 4: Establish Monitoring Procedures

Once critical limits have been created, determine the best way for your operation to check them. Make sure the limits are consistently met. Identify who will monitor them and how often.

Principle 4 Example

At Enrico's, each charbroiled chicken breast is cooked to order. The team decided to check the critical limit by inserting a clean and sanitized thermocouple probe into the thickest part of each chicken breast.

The grill cook must check the temperature of each chicken breast after cooking. Each chicken breast must reach the minimum internal temperature of 165°F (74°C) for 15 seconds.

Principle 5: Identify Corrective Actions

Identify steps that must be taken when a critical limit is not met. These steps should be determined in advance.

Principle 5 Example

If the chicken breast has not reached its critical limit within the 16-minute cook time, the grill cook at Enrico's must keep cooking the chicken breast until it has reached it.

This and all other corrective actions are noted in the temperature log.

Principle 6: Verify That the System Works

Determine if the plan is working as intended. Evaluate it on a regular basis. Use your monitoring charts, records, hazard analysis, etc., and determine if your plan prevents, reduces, or eliminates identified hazards.

Principle 6 Example

Enrico's management team performs HACCP checks once per shift. They make sure that critical limits were met and appropriate corrective actions were taken when needed.

They also check the temperature logs on a weekly basis to identify patterns. This helps to determine if processes or procedures need to be changed. For example, over several weeks they noticed problems toward the end of each week. The chicken breasts often failed to meet the critical limit. The appropriate corrective action was being taken.

Management discovered that Enrico's received chicken shipments from a different supplier on Thursdays. This supplier provided a six-ounce chicken breast. Enrico's chicken specifications listed a four-ounce chicken breast. Management worked with the supplier to ensure they received four-ounce breasts. The receiving procedures were changed to include a weight check.

Principle 7: Establish Procedures for Record Keeping and Documentation

Maintain your HACCP plan and keep all documentation created when developing it. Keep records for the following actions.

- Monitoring activities
- Taking corrective action
- Validating equipment (checking for good working condition)
- Working with suppliers (e.g., shelf-life studies, invoices, specifications, challenge studies, etc.)

Principle 7 Example

Enrico's management team determined that time-temperature logs should be kept for three months. Receiving invoices would be kept for 60 days. The team used this documentation to support and revise their HACCP plan.

Another HACCP Example

The Enrico's example shows one type of HACCP plan. Another plan may look very different when it deals with food that is processed more simply. For example, food that is prepped and served without cooking needs a different approach.

Here is part of the HACCP plan for The Fruit Basket. This fruit-only restaurant is known for its signature item—the Melon Medley salad.

1 Analyzing hazards   The HACCP team at The Fruit Basket decided to look at hazards for the Melon Medley. The salad has fresh watermelon, honeydew, and cantaloupe. The team determined that bacteria are a risk to the fresh-cut melons.

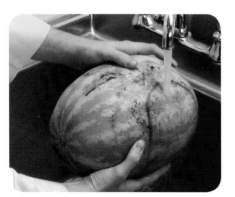

2 Determining CCPs   The melons are prepped, held, and served without cooking. The team determined that preparation and holding are CCPs for the salad. They decided that cleaning and drying the melons' surfaces during prep, as shown in the photo at left, would reduce bacteria. Holding the melon at the right temperature could prevent the bacteria's growth. Receiving was ruled out as a CCP, because the operation purchases melons only from approved suppliers.

3 Establishing critical limits   For the preparation CCP, the team decided the critical limit would be met by washing, scrubbing, and drying whole melons. They created an SOP with techniques for washing the melons. For the holding CCP, they decided that the salad must be held at 41°F (5°C) or lower, because it contained cut melons.

4 Establishing monitoring procedures   The team decided that the operation's team leader should monitor the salad's critical limits. The team leader must observe foodhandlers to make sure they are prepping the melons the right way. Foodhandlers must remove all surface dirt and debris from the washed melons. Then they must cut, mix, and portion the salad into containers. The finished salads are put in the cooler.

The team leader must then monitor the temperature of the held salads to make sure the holding critical limit is met. The internal temperature of the salads must be 41°F (5°C) or lower. It must be checked three times per day. A clean, sanitized, and calibrated probe must be inserted into the salad, as shown in the photo at left.

5 Identifying corrective actions   Sometimes after preparation, the melons still have surface dirt. The team had to determine a corrective action for this. They decided that the action would be to rewash. Then the team leader must approve the melons before they are sliced.

To correct a holding temperature that is higher than 41°F (5°C), the team leader must check the temperature of every Melon Medley in the cooler. Any salad that is above 41°F (5°C) must be thrown away. The team leader must also record all corrective actions in the Manager Daily HACCP Check Sheet.

6 Verifying that the system works   To make sure the system is working right, the team decided that the operation team leader must review the Manager Daily HACCP Check Sheet at the end of each shift. The team leader makes sure that each item was checked and initialed. The team leader also confirms that all corrective actions have been taken and recorded. The Fruit Basket also evaluates the HACCP system quarterly to see if it is working.

7 Establishing procedures for record keeping   Because a foodborne illness associated with fresh produce can take as long as 16 weeks to emerge, the team determined that all HACCP records must be maintained for 16 weeks and kept on file. They also decided that produce suppliers must maintain their records for at least one year.

When a HACCP Plan is Required

A HACCP plan is required when prepping food in the following ways. Always check with your local regulatory authority to see if a variance is also required.

- Smoking food as a method to preserve it (but not to enhance flavor).

- Using food additives or adding components such as vinegar to preserve or alter it so it no longer requires time and temperature control for safety.

- Curing food.

- Custom-processing animals. For example, this may include dressing deer in the operation for personal use.

- Packaging food using reduced-oxygen packaging (ROP) methods. This includes MAP, vacuum-packed, and *sous vide* food. *Clostridium botulinum* and *Listeria monocytogenes* are risks to food packaged in these ways.

- Treat (e.g., pasteurize) juice on-site, and package it for later sale.

- Sprouting seeds or beans.

- Offering live, molluscan shellfish from a display tank.

Apply Your Knowledge

It's the Principle of the Thing

Identify the HACCP principle defined by each statement. Put the number of the principle in the space provided.

Ⓐ _____ Checking to see if critical limits are being met

Ⓑ _____ Keep HACCP plan documents

Ⓒ _____ Assessing risks within the flow of food

Ⓓ _____ Specific places in the flow of food where a hazard can be prevented, eliminated, or reduced to a safe level

Ⓔ _____ Predetermined step taken when a critical limit is not met

Ⓕ _____ Minimum or maximum boundaries that must be met to prevent a hazard

Ⓖ _____ Determining if the HACCP plan is working as intended

① Hazard analysis

② Critical control points

③ Critical limits

④ Monitoring

⑤ Corrective action

⑥ Verification

⑦ Record keeping and documentation

For answers, please turn to page 9.21.

Crisis Management

Putting the food safety principles you have learned into action can help keep food safe in your operation. However, despite your best efforts, a foodborne-illness outbreak may occur. How you respond can make a difference in the outcome.

To deal with a foodborne-illness outbreak, you will need a crisis-management program. Put together a crisis-management team and build a program before you have a problem. This will help you deal with any possible outbreak and recover afterwards.

Building a Successful Crisis-Management Program

A successful crisis-management program has a written plan, as shown in the photo at left. The basic objectives of the plan focus on three parts: preparation, response, and recovery. For each of these parts, the plan must identify the resources needed and the procedures to be followed.

The time to prepare for a crisis is before one happens. There is no "off-the-shelf" disaster plan that works for every operation. Each plan must be tailored to meet the operation's individual needs. A good way to make sure your plan meets your needs is to test the plan once it's complete. The results of the test will help you identify potential gaps or problems.

You need to prepare for many types of crises. Here are some examples.

- Foodborne illness
- Food defense issues
- Product recalls
- Water interruption
- Power outage
- Sewage backup
- Flood

Creating a Crisis-Management Team

To begin, create a crisis-management team, as shown in the photo at left. The size of the team will depend on the size of the operation. However, you should always get management's support. If your operation is large, the team may include representatives from many departments.

- Senior management (president/CEO, etc.)
- Risk management (quality assurance, legal, etc.)
- Public relations
- Operations
- Finance
- Marketing
- Human resources

Smaller operations may include the chef, general manager, and owner/operator. Regardless of size, you should also consider using other resources. External resources include your local regulatory authority and experts from your suppliers and manufacturers.

Preparing for a Foodborne-Illness Outbreak

Preparing for a crisis is the first step in handling one. The greatest threat to your customers is foodborne illness, so you must be ready to deal with an outbreak. Here is one way to prepare.

Food safety program   Make sure the program trains all staff on the policies and procedures that will keep food safe in your operation.

Foodborne-illness incident report form   Develop this form for your operation with legal help. Make sure to include all critical information, similar to the example in the photo at left. There are several questions you may choose to include.

- When and what the customer ate at the operation
- When the customer first became ill
- Medical attention received by the customer
- Other food eaten by the customer

Training   Train staff to fill out incident forms.

Emergency contact list   Make a list with contact information. Include the local regulatory authority, testing labs, and management headquarters personnel.

Crisis-communication plan   The plan must identify the person in charge of a crisis, subject-matter experts, and a media spokesperson.

Responding to a Foodborne-Illness Outbreak

The next step in building a crisis-management plan is including how to respond to a crisis. In a foodborne-illness outbreak, you may be able to avoid a crisis by quickly responding to customer complaints. Here are some things you should consider when responding to an outbreak.

Foodborne-Illness Outbreak Responses

If	Then
A customer calls to report a foodborne illness.	• Take the complaint seriously and express concern. Do **NOT** admit responsibility or accept liability. • Complete the foodborne-illness incident report form, as shown in the photo at left. • Evaluate the complaint to determine if there are similar complaints.
There are similar customer complaints of foodborne illness.	• Contact the crisis-management team. • Identify common food items to determine the potential source of the complaint. • Contact the local health department to assist with the investigation.
The suspected food is still in the operation.	• Isolate the suspected food and identify it to prevent selling it. • If possible, get samples of the suspect food from the customer.
The suspected outbreak is caused by an ill staff member.	• Exclude the suspect staff member from the operation.
The regulatory authority confirms your operation is the source of the outbreak.	• Cooperate with the regulatory authority to resolve the crisis.
The media contacts your operation.	• Follow your crisis-communication plan. Let your spokesperson handle all communication.

Recovering from a Foodborne-Illness Outbreak

The final step in a crisis-management plan is developing procedures for recovering from a crisis. Think about what you need to do to make sure that the operation and the food are safe. This is critical for getting your operation running again. Consider the following in your recovery plan.

- Work with the regulatory authority to resolve issues.

- Clean and sanitize all areas of the operation so the incident does not happen again.

- Throw out all suspect food.

- Investigate to find the cause of the outbreak.

- Establish new procedures or revise existing ones based on the investigation results. This can help to prevent the incident from happening again.

- Develop a plan to reassure customers that the food served in your operation is safe.

Apply Your Knowledge

What's When?

From the list provided, select the activities best associated with each question. Write the letters for the activity after the question.

① What activities should be performed before a foodborne-illness outbreak happens?

_____

② What activities should be performed during a foodborne-illness outbreak?

_____

③ What activities should be performed after a foodborne-illness outbreak?

_____

Ⓐ Implement food safety programs.

Ⓑ Throw out all suspect food.

Ⓒ Take customer complaints seriously. Express regret, but do not accept liability.

Ⓓ Establish new procedures or update old ones based on investigation findings.

Ⓔ Train staff in basic food safety.

Ⓕ Evaluate customer complaints to see if there are other, similar complaints.

Ⓖ Prepare a crisis-management program.

Ⓗ Cooperate with the regulatory authority to resolve the crisis.

Ⓘ Develop a foodborne-illness incident report form.

Ⓙ Fill out a foodborne-illness incident report form.

Ⓚ Refer media representatives to the spokesperson.

Ⓛ Select the members of a crisis-management team.

Ⓜ Get samples of suspect food from the customer.

For answers, please turn to page 9.21.

Chapter Summary

A food safety management system is a group of procedures and practices intended to prevent foodborne illness. The two primary systems are HACCP and active managerial control.

You can use active managerial control to manage food safety risks. This approach focuses on controlling the CDC's most common risk factors responsible for foodborne illness. These are: purchasing food from unsafe sources, failing to cook food adequately, holding food at incorrect temperatures, using contaminated equipment, and practicing poor personal hygiene.

A HACCP system identifies hazards at specific points throughout the flow of food. These hazards can be biological, chemical, or physical. Once identified, hazards can be prevented, eliminated, or reduced to safe levels. An effective HACCP system must be based on a written plan. A HACCP plan is developed following seven basic principles.

A food safety management system can help keep food safe in your operation. But a foodborne-illness outbreak can still happen. The time to prepare for a crisis is before you have one. You need to start with a written plan. The plan must focus on preparation, response, and recovery.

If you must respond to a foodborne-illness complaint, listen to the customer carefully. Express concern, and be sincere. But do not admit responsibility. If there are similar complaints, call your crisis-management team together. Then implement your plan.

Chapter Review Case Study

Now take what you have learned in this chapter and apply it to the following case study.

Maria saw it on the evening news. Another operation in town had a foodborne-illness outbreak. As an owner/ operator, she realized that she needed to do more to keep her place safe. That meant taking charge of food safety in a more formal way than before. It was time to develop an active managerial control plan. While watching the coverage, she decided to make a crisis-management plan too.

She realized that she had to address the four CDC risk factors. Figuring out where cooking, holding, contamination, and personal hygiene posed a danger was the first step. Then Maria started looking at her existing policies and SOPs. She realized that she would probably have to do some extra training. Once the new policies were developed, she would have to monitor them.

Maria listed the different measurements and logs that would be required in her plan. With the plan now complete, she moved on to her risk-management plan. Over the next couple of weeks, she managed to develop a complete plan.

Fortunately, Maria completed both of her plans. Unfortunately, her operation experienced a foodborne-illness outbreak two days later. She had just started to implement her new active managerial control procedures. When the first customer complaint came in, her staff filled out the foodborne-illness incident report. By the time her manager had called her at home, they had four complaints. Maria apologized to the customers. She told them that she would look into the problem. Reviewing the incident reports, she noticed something. All of the customers had eaten the same thing: meatloaf. Maria instructed her staff not to serve any more of it.

It was no surprise when the local health department arrived the next day. Maria worked closely with the authorities. Eventually, they tracked the problem back to one of her newer suppliers. Maria had recently switched to a new

Continued on the next page ▶

► *Continued from previous page*

meat supplier because they were cheaper. It turned out that the supplier had inadequate cold-holding facilities. The meat had been time-temperature abused before it ever arrived at her operation. The bad food was thrown out immediately. Maria continued to work with the health department. Together they established new guidelines for selecting suppliers.

① What did Maria do right?

_____

_____

_____

_____

_____

② What could Maria have done differently?

_____

_____

_____

_____

_____

For answers, please turn to page 9.21.

Study Questions

Circle the best answer to each question below.

① **The temperature of a roast is checked to see if it has met its critical limit of 145°F (63°C) for 4 minutes. This is an example of which HACCP principle?**

A Verification

B Monitoring

C Record keeping

D Hazard analysis

② **The temperature of a pot of beef stew is checked during holding. The stew has not met the critical limit and is thrown out according to house policy. Throwing out the stew is an example of which HACCP principle?**

A Corrective action

B Hazard analysis

C Verification

D Monitoring

③ **The CDC has determined five common risk factors for foodborne illness. They are: purchasing food from unsafe sources, failing to cook food adequately, holding food at incorrect temperatures, practicing poor personal hygiene, and using**

A imported supplies.

B incorrect shellstock tags.

C unapproved chemicals.

D contaminated equipment.

④ **What is the first step in developing a HACCP plan?**

A Identify corrective actions.

B Conduct a hazard analysis.

C Establish monitoring procedures.

D Determine critical control points.

⑤ **A food safety management system is a group of _____ for preventing foodborne illness.**

A managers and customers

B measurements and graphs

C procedures and practices

D detergents and sanitizers

Continued on the next page ▶

▶ *Continued from previous page*

⑥ **What is the purpose of a food safety management system?**

A   To keep all areas of the facility clean and pest free

B   To identify, tag, and repair faulty equipment within the facility

C   To identify, document, and use the correct methods for receiving food

D   To identify and control possible hazards in the flow of food

⑦ **A chef sanitized a thermometer probe and then checked the temperature of minestrone soup being held in a hot-holding unit. The temperature was 120°F (49°C), which did not meet the operation's critical limit of 135°F (57°C). The chef recorded the temperature in the log and reheated the soup to 165°F (74°C) for 15 seconds within two hours. Which was the corrective action?**

A   Reheating the soup

B   Checking the critical limit

C   Sanitizing the thermometer probe

D   Recording the temperature in the log

⑧ **An operation that wants to smoke food as a method of preservation must have a(n)**

A   current organization chart.

B   crisis-management plan.

C   HACCP plan.

D   MSDS.

⑨ **What is the third step in active managerial control?**

A   File the documentation in case of a crisis.

B   Monitor the policies and procedures.

C   Revise the policies and procedures.

D   Determine staffing needs.

⑩ **Which is an example of when a HACCP plan is required?**

A   Serving smoked meat on a metal platter

B   Serving chili made from a family recipe

C   Serving wild game with cream sauce

D   Serving raw oysters from a display tank

For answers, please turn to page 9.21.

Answers

9.12   It's the Principle of the Thing

- Ⓐ 4
- Ⓔ 5
- Ⓑ 7
- Ⓕ 3
- Ⓒ 1
- Ⓖ 6
- Ⓓ 2

9.16   What's When?

- ① A, E, G, I, L
- ② C, F, H, J, K, M
- ③ B, D

9.17   Chapter Review Case Study

① Here is what Maria did right.

- Maria recognized the need for both food safety management and crisis-management systems. The time to build these programs is before a problem happens.

- Maria recognized the need to retrain her staff on new procedures and policies. She understands the need to monitor her new procedures to make sure that they keep food safe.

- When the customer complaints started, the staff filled out the foodborne-illness incident report form.

- Maria was not defensive and apologized to the customers.

- She reviewed the incident reports and looked for a common link. In this case, it was the meatloaf.

- Maria worked with the authorities.

- She threw out the contaminated food.

- Maria worked to correct the source of the problem when she created new supplier-selection guidelines.

② Here is what Maria could have done differently.

- When she developed her active managerial control plan, Maria missed one of the five CDC risk factors. She did not include purchasing food from unsafe sources. She should have made a list of criteria for choosing approved suppliers, rather than using only the cheapest.

9.19   Study Questions

- ① B
- ⑤ C
- ⑨ B
- ② A
- ⑥ D
- ⑩ D
- ③ D
- ⑦ A
- ④ B
- ⑧ C

10

Sanitary Facilities and Equipment

In the News

Diner Closes to Fix Code Violations

A local diner voluntarily closed last week after a city health inspector found many food safety violations. According to the inspector, the violations could have led to a foodborne-illness outbreak.

All the violations came from the building itself rather than the operation's foodhandling practices. The walls and ceiling in the storage area had gaps and cracks that needed to be sealed. The lid and interior of the ice machine needed to be repaired and cleaned. Lights over the food-prep areas were missing light shields. And perhaps most important, the ice machine had a cross-connection between the drain line and the drinkable water supply. This could have allowed pathogens to flow back into the machine and contaminate the ice.

These problems have been fixed, and the diner has reopened. City health officials say the restaurant is now up to code.

You Can Prevent This

In the story above, even good cleaning practices could not make up for sanitation issues in the building. To make sure your facility is safe for foodservice, you should know the following.

- How a kitchen layout can affect food safety

- How to pick materials and equipment that are safe for use in foodservice operations

- How to install and maintain equipment

- How to avoid food safety hazards caused by utilities

- How to maintain your facility

Concepts from Earlier Chapters

Before reading this chapter, remember this concept.

Cross-contamination Transfer of pathogens from one surface or food to another.

Designing a Sanitary Operation

As you learned in chapter 9, sanitary facilities and equipment are the basic requirements of a food safety system. Many things affect how easy or difficult it is to clean and sanitize your facility. One of the most important is how it is designed.

Facility Design

A facility should be designed so it will keep food safe and can be cleaned quickly and effectively. A well-designed facility has the following features.

Good workflow   The workflow should keep food out of the temperature danger zone as much as possible and limit the number of times food is handled. For example, storage areas should be near the receiving area to prevent delays in storing food. Prep tables should be near coolers and freezers for the same reason. A good layout will also encourage good personal hygiene practices.

The drawing below shows a well-designed facility.

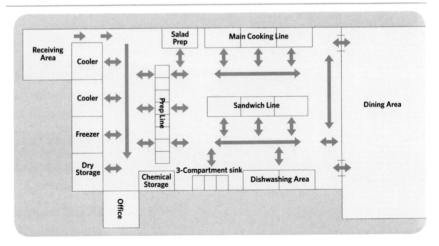

Arrows indicate normal work flow.

Reduction of cross-contamination   Place equipment to prevent splashing or spillage from one piece of equipment onto another. For example, it is not a good practice to place the dirty-utensil table next to the salad-prep sink.

Accessibility for cleaning   Hard-to-reach areas are less likely to be cleaned. A well-planned layout makes it easier for staff to clean the facility and equipment.

Design Review

Before starting any new construction or a large remodeling project, check with your local regulatory authority. You may need approval for your design plan. Even if you don't need approval, you should ask your regulatory authority to review the plan. A review like the one in the photo at left can have several benefits.

- It ensures that the design meets regulatory requirements.

- It ensures a safe flow of food.

- It may save time and money.

Additionally, these authorities provide information on what is necessary for good sanitation.

How This Relates to Me

When an operation is to be remodeled or new construction is planned, does your regulatory authority require approval of the design plan?

Yes _____   No _____

If yes, what authorities must approve the plan?

_____

_____

Apply Your Knowledge

Can It Improve Food Safety?

Decide if each situation is likely to improve food safety. Explain why or why not.

① In the plans for a new family restaurant, the drain board for clean items in a dishwashing area is next to a prep table.

_____

_____

② The plan for a new school cafeteria has the freezers and coolers next to the receiving area.

_____

_____

_____

For answers, please turn to page 10.19.

Interior Requirements for a Sanitary Operation

The materials, equipment, and utilities in your operation play a part in keeping food safe. Given the opportunity, you should choose these items with food safety in mind and based on the criteria below. But choosing the right items for your operation is only half of the job. You also must maintain them to ensure food safety.

Material Selection for Interior Surfaces

When choosing flooring, wall and ceiling materials, and doors, the most important factor is how easy the surfaces are to clean and maintain. They should be replaced if damaged or worn.

Floors

Flooring must be smooth, nonabsorbent, easy to clean, and durable. The kitchen in the photo at left has this type of flooring. It should be used in the following areas.

- Food prep
- Food storage
- Dishwashing
- Walk-in coolers
- Dressing and locker rooms
- Restrooms

In addition, floors should have coving. Coving is a curved, sealed edge between a floor and a wall, as shown in the photo at left. It gets rid of sharp corners or gaps that are hard to clean. Coving should be glued tightly to the wall to get rid of hiding places for insects. This also protects the wall from moisture.

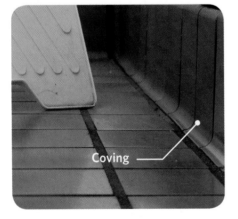

Coving

Walls, Ceilings, and Doors

The materials for your facility's walls, ceilings, and doors must be smooth, nonabsorbent, durable, and easy to clean. Light colors are recommended for walls and ceilings. Walls should also be able to withstand repeated washing.

Equipment Selection

Equipment must meet certain standards that depend on whether or not the equipment's surfaces come in direct contact with food. The equipment you buy must meet the following requirements.

Food-contact surfaces   Food-contact surfaces must have these traits.

- Safe for contact with food
- Nonabsorbent, smooth, and corrosion resistant, as shown in the photo at left
- Easy to clean and maintain
- Durable—stands up to heavy use and repeated cleaning
- Resistant to damage such as pitting, chipping, crazing (spider cracks), scratching, scoring, distortion, and decomposition

Nonfood-contact surfaces   These surfaces aren't designed for direct contact with food, but food may splash or spill onto them. They must have these traits.

- Nonabsorbent, smooth, and corrosion resistant
- Easy to clean and maintain
- Free of unnecessary ledges, projections, and crevices

Fortunately, there are organizations to help with the task of choosing equipment. Look for the NSF mark or the UL classified or UL EPH listed marks on foodservice equipment. Only use equipment designed for use in a foodservice operation.

NSF creates standards for foodservice equipment. It also certifies equipment. The NSF mark means an item has been evaluated, tested, and certified by NSF as meeting its food-equipment standards.

Underwriters Laboratories (UL) provides classification listings for equipment that meets ANSI/NSF standards.

UL also certifies items that meet its own standards for environmental and public health (EPH). Equipment that meets UL EPH standards is also acceptable for foodservice use. This is shown by the UL EPH Listed mark.

Installing and Maintaining Equipment

Stationary equipment should be easy to clean and easy to clean around. How you install it can make a big difference. In the photo at left, the dishwasher is installed so the floor can be cleaned easily.

When installing equipment, follow manufacturers' recommendations. Also, check with your regulatory authority for requirements. In general, stationary equipment should be installed as follows.

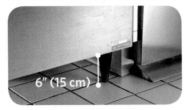

Floor-mounted equipment   Put floor-mounted equipment on legs at least six inches (15 centimeters) high. Another option is to seal it to a masonry base.

Tabletop equipment   Put tabletop equipment on legs at least four inches (10 centimeters) high. Or seal it to the countertop.

Gaps   Seal any gaps between equipment and surrounding countertops and walls.

Once equipment is installed, it must be maintained regularly. Only qualified people should do this work. Also, set up a maintenance schedule with your supplier or manufacturer. Check equipment regularly to be sure it is working right.

Dishwashing Machines

Dishwashers vary by size, style, and sanitizing method. For example, some sanitize with very hot water. Others use a chemical solution.

Consider these guidelines when selecting and installing dishwashers.

Settings   Information about the right settings should be posted on the machine. It should show the right water temperature, conveyor speed, and water pressure. The sign in the photo at left is an example.

Thermometer   The machine's thermometer should be in a place where it is easy to read. It should show the temperature in increments of no greater than 2°F (1°C).

Handwashing Stations

Handwashing stations should be put in areas that make it easy for staff to wash their hands often. These stations are required in restrooms and in areas used for food prep, service, and dishwashing. Make sure these stations work right and are well stocked and maintained.

A handwashing station must have the following items.

Hot and cold running water   Hot and cold water should be supplied through a mixing valve or combination faucet. The hot water temperature should reach at least 100°F (38°C).

Soap   The soap can be liquid, bar, or powder.

A way to dry hands   Most local codes require disposable paper towels. A hand dryer may also be used as a backup. It must dry hands using warm air or room-temperature air delivered at high velocity.

If continuous-cloth towel systems are allowed for your operation, only use them if the unit is working correctly. Also, make sure the towel rolls are checked and changed regularly.

Do **NOT** use common cloth towels. They can transfer dirt and pathogens from one person's hands to another's.

Garbage container   Garbage containers are required if disposable paper towels are used.

Signage   There must be a sign that tells employees to wash hands before returning to work. The message should be in all languages used by employees in the operation.

Check your local regulatory requirements.

Utilities and Building Systems

An operation uses many utilities and building systems. Utilities include water, electricity, gas, sewage, and garbage disposal. Building systems include plumbing, lighting, and ventilation. There must be enough utilities to meet the needs of the operation. In addition, the utilities and systems must work correctly. If they do not, the risk of contamination is greater.

Water and Plumbing

Water is used for dishwashing, cleaning, cooking, and drinking in an operation. Having safe water is critical. When water is safe to drink, it is called potable. It may come from these sources.

- Approved public water mains

- Private water sources that are regularly tested and maintained

- Closed, portable water containers

- Water transport vehicles

If your operation uses a private water supply, such as a well, you must make sure the water is safe to use. Check with your local regulatory authority for information on inspections, testing, and other requirements. Generally, you should test private water systems at least once a year, as shown in the photo at left. Keep these test reports on file.

Regardless of where your water comes from, you should know how to prevent plumbing issues that can affect food safety.

Installation and maintenance   Plumbing that is not installed or maintained the right way can allow potable and unsafe water to be mixed. This can cause foodborne-illness outbreaks. Have only licensed plumbers work on the plumbing in your operation.

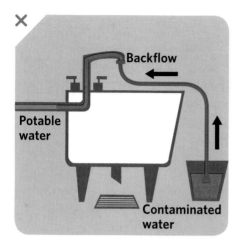

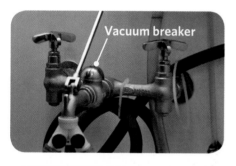

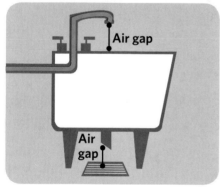

Cross-connection   The greatest challenge to water safety comes from cross-connections. A cross-connection is a physical link between safe water and dirty water, which can come from drains, sewers, or other wastewater sources.

A cross-connection is dangerous because it can let backflow occur. Backflow is the reverse flow of contaminants through a cross-connection into a potable water supply. Backflow can happen when the pressure in a potable water supply drops below the pressure of dirty water. The pressure difference can pull the dirty water into the safe water supply. The graphic at left shows both a cross-connection and backflow.

A running faucet below the flood rim of a sink is an example of a cross-connection that can lead to backflow. A running hose in a mop bucket is another example.

Backflow prevention   The best way to prevent backflow is to avoid creating a cross-connection. Do NOT attach a hose to a faucet unless a backflow prevention device, such as the vacuum breaker shown in the photo at left, is attached. Threaded faucets and connections between two piping systems must have a vacuum breaker or other approved backflow prevention devices. Even if these devices are installed, NEVER create a cross-connection. This way, if the device breaks, your water supply will not be endangered.

The only sure way to prevent a backflow is to create an air gap. An air gap is an air space that separates a water supply outlet from a potentially contaminated source. A sink that is correctly designed and installed usually has two air gaps, as shown in the graphic at left. One is between the faucet and the flood rim of the sink. The other is between the drainpipe of the sink and the floor drain of the operation.

Grease condensation   A buildup of grease in pipes is another common problem in plumbing systems. Grease traps are often installed to prevent a grease buildup from blocking the drain. If used, they should be put in by a licensed plumber and be easy to access. Also, make sure they are cleaned regularly following the manufacturer's recommendations. If the traps are not cleaned often enough or correctly, dirty water can back up. This backup could lead to odors and contamination.

Overhead leaks   Overhead wastewater pipes or fire-safety sprinkler systems can leak and cause contamination. Even overhead pipes carrying potable water can be a problem. This is because water can condense on the pipes and drip onto food. Check all pipes regularly to make sure they are in good shape and do not leak.

Sewer

Sewage and wastewater are contaminated with pathogens, dirt, and chemicals. You must prevent them from contaminating food or food-contact surfaces. If raw sewage backs up in your operation, close the affected area right away. Then fix the problem and thoroughly clean the area.

A facility's drain system must be able to handle all wastewater. Areas with a lot of water should have floor drains, as shown in the photo at left.

Lighting

Good lighting has many benefits. It helps improve work habits and makes it easier to clean things. It also provides a safer environment. Lighting requirements are usually measured in units called foot-candles or lux. The following table shows the lighting required in various areas of an operation.

Minimum Lighting Intensity Requirements	
Minimum Lighting Intensity	**Area**
50 foot-candles (540 lux)	• Prep areas
20 foot-candles (215 lux)	• Handwashing or dishwashing areas • Buffets and salad bars • Displays for produce or packaged food • Utensil-storage areas • Wait stations • Restrooms • Inside some equipment (e.g., reach-in coolers)
10 foot-candles (108 lux)	• Inside walk-in coolers and freezer units • Dry-storage areas • Dining rooms (for cleaning purposes)

All lights should have shatter-resistant lightbulbs or protective covers. These products prevent broken glass from contaminating food or food-contact surfaces.

Ventilation

Ventilation improves the air inside an operation. It removes odors, gases, grease, dirt, and mold. These things can cause contamination. If ventilation is poor, grease and condensation will build up on walls and ceilings.

Ventilation must be designed so that grease and condensation from hoods, fans, and ductwork do not drip onto food or equipment. Hood filters or grease extractors must be tight fitting but easy to take off, as shown in the photo at left. Make sure they are cleaned on a regular basis. Also, have a professional expert clean the hood and ductwork periodically. As a manager, you are responsible for making sure the ventilation system meets all regulatory requirements.

Garbage

Garbage can attract pests and contaminate food, equipment, and utensils if not handled the right way. To control hazards from garbage, consider the following.

Garbage removal   Garbage should be removed from prep areas as quickly as possible to prevent odors, pests, and possible contamination.

Cleaning of containers   Clean the inside and outside of garbage containers frequently. This will help reduce odors and pests. Do NOT clean garbage containers near prep or food-storage areas. The employee in the photo at left is cleaning a garbage container in an area that will not put food safety at risk.

Indoor containers   Containers must be leak proof, waterproof, and pest proof. They also should be easy to clean.

Liners for containers   Line garbage containers with plastic bags or wet-strength paper bags.

Outdoor containers   Place garbage containers on a smooth, nonabsorbent surface, such as the one in the photo at left. Make sure they have tight-fitting lids and are kept covered at all times. Keep their drain plugs in place.

Maintaining the Facility

Poor maintenance can cause food safety problems in your operation. To prevent them, do the following.

- Clean the operation on a regular basis.

- Make sure all building systems work and are checked regularly.

- Make sure the building is sound. There should be no leaks, holes, or cracks in the floors, foundation, ceilings, or windows. In the photo at left, the maintenance worker is filling the crack to keep pests out.

- Control pests.

- Maintain the outside of the building and property, including patios and parking lots.

Apply Your Knowledge

Which Materials Should You Use?

Which of these traits are required for materials inside your operation? Pick all the traits needed for each area. (A trait may be used more than once.)

① _____ Floors   Ⓐ Pitted

② _____ Walls   Ⓑ Smooth

③ _____ Ceilings   Ⓒ Durable

④ _____ Doors   Ⓓ Absorbent

                                    Ⓔ Dark colored

                                    Ⓕ Light colored

                                    Ⓖ Easy to clean

                                    Ⓗ Nonabsorbent

How Bright Is It?

Identify the minimum lighting-intensity requirement for each area. Write the letter in the space provided.

① _____ Dry-storage areas   Ⓐ 50 foot-candles (540 lux)

② _____ Prep areas   Ⓑ 20 foot-candles (215 lux)

③ _____ Handwashing areas   Ⓒ 10 foot-candles (108 lux)

④ _____ Walk-in refrigerators

⑤ _____ Dishwashing areas

Which Traits Are Best for Foodservice Equipment?

Which of these traits should you choose for foodservice equipment? Pick all the traits that apply for each type of surface. (A trait may be used more than once.)

① _____ Food-contact surfaces

② _____ Nonfood-contact surfaces

Ⓐ Safe for contact with food

Ⓑ Smooth

Ⓒ Durable

Ⓓ Absorbent

Ⓔ Light colored

Ⓕ Dark colored

Ⓖ Easy to clean

Ⓗ Rough textured

Ⓘ Nonabsorbent

Ⓙ Easy to maintain

Ⓚ Corrosion resistant

Ⓛ Damage resistant

Ⓜ Free of unnecessary ledges, projections, and crevices

Ⓝ Free of unnecessary features, decoration, and controls

Which Marks Certify Equipment for Foodservice Use?

Which of these marks mean equipment is safe to use in a foodservice operation? Place a ✔ next to each one.

① _____

② _____

③ _____

④ _____

⑤ _____

⑥ _____

For answers, please turn to page 10.19.

How Should You Install and Maintain Stationary Equipment?

How should each of the following types of stationary equipment be installed and maintained? Pick all the options that apply for each type of equipment. (An option may be used more than once.)

1 _____ Floor-mounted equipment

2 _____ Tabletop equipment

Ⓐ Used regularly

Ⓑ Inspected regularly

Ⓒ Maintained regularly

Ⓓ Sealed to a countertop

Ⓔ Sealed to a masonry base

Ⓕ Sealed gaps between equipment and surrounding surfaces

Ⓖ Mounted on legs at least four inches (10 centimeters) high

Ⓗ Mounted on legs at least six inches (15 centimeters) high

Ⓘ Mounted on legs at least eight inches (20 centimeters) high

What's Missing?

The handwashing station is missing three items. What are they?

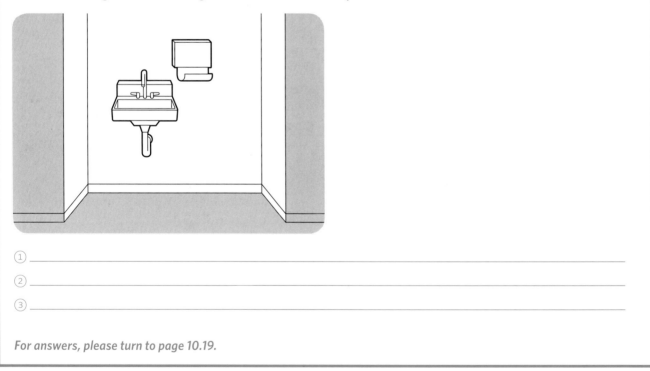

1 _____

2 _____

3 _____

For answers, please turn to page 10.19.

Chapter Summary

A facility should be designed so it will keep food safe and can be cleaned quickly and effectively. Good design and selection of floors, walls, ceilings, and doors in a facility will help keep food safe. Choose materials based on how easy they are to clean and maintain. Use equipment that is easy to clean and to clean around. Food-contact surfaces must be corrosion resistant, nonabsorbent, and smooth. They must also be resistant to pitting and scratching.

Handwashing stations must be fully stocked and maintained. They are required in prep areas, service areas, dishwashing areas, and restrooms. They must also have hot and cold running water, soap, a way to dry hands, a garbage container, and signs reminding employees to wash their hands.

Utilities and building systems must work correctly. Otherwise, they can cause contamination. Plumbing that is not installed or maintained correctly can make food unsafe. Only licensed plumbers should work on plumbing systems. The greatest water safety risk comes from cross-connections. These are physical links between safe and dirty water. These can come from drains, sewers, and other wastewater sources. Use vacuum breakers and air gaps to prevent backflow. If sewage backs up, you must close the area right away. Then fix the problem, and thoroughly clean the area.

Good lighting helps to improve employee work habits and makes cleaning easier. It also creates a safer work environment. Good ventilation improves the indoor air quality of the facility. It removes odors, gases, grease, dirt, and mold. Design ventilation so hoods, fans, guards, and ductwork do not drip onto food or equipment. Clean hood filters and grease extractors regularly. Garbage containers must be leak proof, waterproof, and pest proof. Clean the inside and outside of garbage containers frequently. Remove garbage from prep areas as soon as possible.

Chapter Review Activities

Now take what you have learned in this chapter and apply it to the following case studies.

① Charlotte is the owner of The Sandbar Café, a small storefront restaurant. She is planning to rent the space next door, take down some walls, and expand her kitchen and dining room. To help with this, she hired an architect to draw up some plans. She feels lucky to have found this architect because he has designed many foodservice facilities. When Charlotte asked him about getting permits for building, the architect said the contractor will take care of it. What should Charlotte do before she begins construction?

_____

_____

_____

② Julio is the assistant manager of the employee cafeteria at The Triple Z Company. When he came in this morning, he found that raw sewage had backed up through the floor drain near the freezers. How should he handle this problem?

_____

_____

_____

Continued on the next page ▶

► *Continued from previous page*

③ Anita is the new manager at Delbert's, an independent quick-service restaurant. On her first day, she noticed several things about the facility that needed attention. One was the buildup of grease on the kitchen walls. What should she do to fix this problem and make sure it does not happen again?

_____

_____

_____

④ Jon is the manager of Home Sweet Home, a family restaurant. Since he started work six months ago, there have been some problems with the plumbing at the dishwashing station. First, the sinks have been draining more and more slowly. Then, when the dishwashing machine finished a cycle earlier today, dirty water backed up in the sinks located nearby. What can Jon do to fix these problems?

_____

_____

_____

For answers, please turn to page 10.20.

Study Questions

Circle the best answer to each question below.

① **Generally, operations that use a private water source, such as a well, must have it tested at least**

A every year.

B every 2 years.

C every 3 years.

D every 5 years.

② **What is the only completely reliable method for preventing backflow?**

A Air gap

B Ball valve

C Vacuum breaker

D Cross-connection

③ **When installing tabletop equipment on legs, the space between the base of the equipment and the tabletop must be at least**

A 1 inch (3 centimeters).

B 2 inches (5 centimeters).

C 4 inches (10 centimeters).

D 6 inches (15 centimeters).

④ **How hot should the hot water at a handwashing station get?**

A At least 70°F (21°C)

B At least 100°F (38°C)

C At least 130°F (54°C)

D At least 160°F (71°C)

⑤ **Foodservice equipment that has been certified as meeting certain standards may be stamped with the _____ mark.**

A FDA

B NSF

C USDA

D USPS

Continued on the next page ▶

► *Continued from previous page*

⑥ **What information should be posted on or near a dishwasher?**

A Water temperature, conveyor speed, and water pressure

B Load start time, estimated load completion time, and initials of person who started the load

C Water temperature in 4°F (2°C) increments and clearance space between the machine and the floor

D Maintenance records including service dates, work done, and name of the company that did the work

⑦ **To keep food from being contaminated by lighting, use**

A shields on heat lamps.

B signage next to lights in food-contact areas.

C fluorescent and other energy-efficient lightbulbs.

D lighting that meets the minimum intensity requirements.

⑧ **Which is a source of potable water?**

A Collected rain water

B Gray-water collection tanks

C Untested private water sources

D Water transport vehicles

⑨ **Outdoor garbage containers should be**

A labeled with collection times.

B kept covered with tight-fitting lids.

C kept away from customer parking areas.

D lined with plastic or wet-strength paper.

⑩ **What is a cross-connection?**

A Threaded faucet

B Device that prevents a vacuum

C Valve that mixes hot and cold water

D Link between sources of safe and dirty water

⑪ **Backflow is when contaminated water**

A flows in pipes behind a wall.

B backs up in a drain because of grease condensation.

C goes down the drain in a counter-clockwise direction.

D flows in reverse because of water pressure.

For answers, please turn to page 10.20.

Answers

10.3 Can It Improve Food Safety?

① No. Food from the prep table could contaminate clean items or the drain board.

② Yes. Putting the freezers and coolers next to the receiving area can reduce the time that food spends in the temperature danger zone.

10.12 Which Materials Should You Use?

① B, C, G, H

② B, C, F, G, H

③ B, C, F, G, H

④ B, C, G, H

10.12 How Bright Is It?

① C

② A

③ B

④ C

⑤ B

10.13 Which Traits Are Best for Foodservice Equipment?

① A, B, C, G, I, J, K, L

② B, G, I, J, K, M

10.13 Which Marks Certify Equipment for Foodservice Use?

2, 4, and 5 should be marked.

10.14 How Should You Install and Maintain Stationary Equipment?

① B, C, E, F, H

② B, C, D, F, G

10.14 What's Missing?

① Soap

② Sign stating that employees must wash hands before returning to work

③ Garbage container for used paper towels

Continued on the next page ▶

▶ *Continued from previous page*

10.15 Chapter Review Activities

① Charlotte should check with her local regulatory authority. Even if the regulatory authority does not require the plans to be approved, a review has many benefits. These include ensuring a safe flow of food; ensuring that the design and layout meet regulatory requirements; and saving time and money.

② Close the area, fix the problem, and thoroughly clean the area.

③ Have the walls washed regularly. Have the ventilation system cleaned regularly. Have a professional company clean the hood and ductwork periodically.

④ If the sinks backed up, this indicates that they do not have an air gap in their drain. This must be fixed. If grease traps are installed, have them cleaned regularly. If grease traps are not installed, have them installed by a licensed plumber.

10.17 Study Questions

① A

② A

③ C

④ B

⑤ B

⑥ A

⑦ A

⑧ D

⑨ B

⑩ D

⑪ D

Notes

11

Cleaning and Sanitizing

In the News

Salmonella Traced to Tea Machine

The local health department has linked more than a dozen cases of salmonellosis to Oak Grove Dining and Banquets. The contamination was traced to its iced tea urn. The local health department said *Salmonella* spp. was found in the spigot of the urn. In response, Oak Grove's owner replaced the urn. He also closed the facility for several days to clean and sanitize the kitchen.

Health department officials said that people are often surprised to hear that tea equipment can be a source of contamination. They recommended equipment that comes in contact with tea be cleaned and sanitized at least once a day. They also said that spigots should be replaced at the end of the day with clean and sanitized ones.

You Can Prevent This

In the story you just read, customers got sick because the operation did not clean and sanitize its tea urn the right way. Cleaning and sanitizing food-contact surfaces can help you avoid foodborne-illness outbreaks. To do it right, you need to know about the following topics.

- How and when to clean and sanitize surfaces
- The different methods of sanitizing and how to make sure they are effective
- How to wash items in a dishwasher or a three-compartment sink and then store them
- How to use and store cleaning tools and supplies
- How to develop a cleaning program

Concepts from Earlier Chapters

Before reading this chapter, remember these concepts and facts.

Pathogens   Certain viruses, parasites, fungi, and bacteria that can cause illness.

Food-contact surface   Any surface that touches food, such as knives, stockpots, and cutting boards.

Cleaning and Sanitizing

Food can easily be contaminated if you don't keep your facility and equipment clean and sanitized. Surfaces that touch food must be cleaned and sanitized the right way and at the right times. Cleaning includes using the right type of cleaner for a job. Sanitizing involves using a method that works for your operation and following the right steps to make sure it is effective.

How and When to Clean and Sanitize

Cleaning removes food and other dirt from a surface. Sanitizing reduces pathogens on a surface to safe levels.

All surfaces must be cleaned and rinsed. This includes walls, storage shelves, and garbage containers. However, any surface that touches food, such as knives, stockpots, and cutting boards, must be cleaned *and* sanitized.

How to clean and sanitize   To clean and sanitize a surface, follow these steps.

❶ Clean the surface.

❷ Rinse the surface.

❸ Sanitize the surface.

❹ Allow the surface to air-dry.

When to clean and sanitize All food-contact surfaces need to be cleaned and sanitized at these times.

- After they are used

- Before foodhandlers start working with a different type of food

- Any time foodhandlers are interrupted during a task and the items being used may have been contaminated

- After four hours if items are in constant use

Cleaners

Cleaners are chemicals that remove food, dirt, rust, stains, minerals, and other deposits. Cleaners must be stable, noncorrosive, and safe to use. Ask your supplier to help you pick cleaners that meet your needs.

Guidelines for Cleaners

To use cleaners correctly, follow these guidelines.

- Follow manufacturers' instructions carefully, as the manager in the photo at left is doing. If not used the right way, cleaners may not work and can even be dangerous.

- NEVER combine cleaners. Combining ammonia and chlorine bleach, for example, produces chlorine gas. This gas can be fatal.

- Do NOT use one type of detergent in place of another unless the intended use is the same. For example, if you use dishwasher detergent to wash dishes by hand, you can burn your skin. Check the label for the intended use.

Types of Cleaners

Cleaners are divided into the following four groups.

Detergents The detergent you use will depend on your task.

- General-purpose detergents remove fresh dirt from floors, walls, ceilings, prep surfaces, and most equipment and utensils.

- Heavy-duty detergents remove wax, aged or dried dirt, and baked-on grease. Dishwasher detergents are an example.

Degreasers Degreasers have ingredients for dissolving grease. They work well where grease has been burned on. This includes grill backsplashes, oven doors, and range hoods.

Delimers Delimers are acid cleaners used on mineral deposits and dirt that other cleaners can't remove. Delimers are often used on steam tables and dishwashers, as shown in the photo at left.

Abrasive cleaners Abrasive cleaners have a scouring agent that helps scrub hard-to-remove dirt. They are used to remove baked-on food. Be aware that they can scratch surfaces.

Sanitizing

Food-contact surfaces must be sanitized after they have been cleaned and rinsed. This can be done by using heat or chemicals.

Heat Sanitizing

One way to sanitize items is to soak them in hot water. For this method to work, the water must be at least 171°F (77°C). The items must be soaked for 30 seconds. You may need a heating device to keep the water hot enough for sanitizing. Be sure to check the water with a thermometer.

Another way to sanitize items is to run them through a high-temperature dishwasher. You can check the water temperature in these machines in two ways.

- Attach a temperature-sensitive label or tape to an item and run it through the dishwasher, as shown in the photo at left.

- Place a high-temperature thermometer in a dish rack and run it through the dishwasher.

Chemical Sanitizing

Tableware, utensils, and equipment can be sanitized by soaking them in sanitizing solution. Or you can rinse, swab, or spray them with sanitizing solution, as shown in the photo at left.

Three common types of chemical sanitizers are chlorine, iodine, and quaternary ammonium compounds, or quats. Chemical sanitizers are regulated by state and federal environmental protection agencies (EPAs). For a list of approved sanitizers, check the Code of Federal Regulations (CFR) 40CFR180.940—"Food-Contact Surface Sanitizing Solutions." For recommendations, check with your local regulatory authority.

In some cases, you can use detergent-sanitizer blends to sanitize. Operations that have two-compartment sinks often use these. If you use a detergent-sanitizer blend, use it once to clean. Then use it a second time to sanitize.

Sanitizer Effectiveness

The table at the bottom of this page gives general guidelines for the effective use of chlorine, iodine, and quats.

Concentration   Sanitizer solution is a mix of chemical sanitizer and water. The concentration of this mix—the amount of sanitizer to water—is critical. Too much water may make the solution weak and useless. Too much sanitizer may make the solution too strong and unsafe. It could also leave a bad taste on items or corrode metal.

Concentration is measured in parts per million (ppm). To check the concentration of a sanitizer solution, use a test kit, shown in the photo at left. Make sure it is made for the sanitizer being used. These kits are usually available from the chemical manufacturer or supplier.

Hard water, food bits, and leftover detergent can reduce the solution's effectiveness. Change the solution when it looks dirty or its concentration is too low. Check the concentration often.

Temperature   The water in sanitizing solution must be the right temperature. Follow manufacturers' recommendations.

Contact time   For a sanitizer solution to kill pathogens, it must make contact with the object being sanitized, such as the baine in the photo at left, for a specific amount of time.

Water hardness   Water hardness can affect how well a sanitizer works. Water hardness is the amount of minerals in your water. Find out what your water hardness is from your municipality. Then work with your supplier to identify the right amount of sanitizer to use for your water.

pH   Water pH can also affect a sanitizer. Find out what the pH of your water is from your municipality. Then work with your supplier to find out the right amount of sanitizer to use for your water.

General Guidelines for the Effective Use of Chlorine, Iodine, and Quats

	Chlorine		Iodine	Quats
Water temperature	≥100˚F (38˚C)	≥75˚F (24˚C)	68˚F (20˚C)	75˚F (24˚C)
Water pH	≤10	≤8	≤5 or as per manufacturer's recommendation	As per manufacturer's recommendation
Water hardness	As per manufacturer's recommendation		As per manufacturer's recommendation	≤500 ppm or as per manufacturer's recommendation
Sanitizer concentration	50-99 ppm	50-99 ppm	12.5-25 ppm	As per manufacturer's recommendation
Sanitizer contact time	≥7 sec	≥7 sec	≥30 sec	≥30 sec

Apply Your Knowledge

Take the Right Steps

Put the steps for cleaning and sanitizing in order by writing the number of the step in the space provided.

Ⓐ _____ Sanitize the surface.

Ⓑ _____ Clean the surface.

Ⓒ _____ Allow the surface to air-dry.

Ⓓ _____ Rinse the surface.

To Sanitize or Not to Sanitize

Place a ✔ next to each situation that requires the foodhandler to clean and sanitize the item being used.

① _____ Jorge has used the same knife to shuck oysters for two hours.

② _____ Bill finishes deboning chicken and wants to use the same cutting board to fillet fish.

③ _____ Mary returns to the slicer to continue slicing ham after being called away to help with the lunch rush.

④ _____ Maria has been slicing cheese on the same slicer from 8:00 a.m. to 12:00 p.m.

Which Cleaner Should You Use?

Match the best cleaner for the job. Write a letter next to each job. Each letter should be used once.

① _____ Remove bits of baked cheese from a pot.   Ⓐ Detergent

② _____ Clean a grill backsplash.   Ⓑ Degreaser

③ _____ Remove mineral deposits from a steam table.   Ⓒ Delimer

④ _____ Wash a kitchen wall.   Ⓓ Abrasive cleaner

Was It Sanitized?

Place a ✔ next to the correct answer for each question. For all situations, assume water hardness and pH are at the right level.

① Lee mixed a quats sanitizer with 75°F (24°C) water. A test kit showed the concentration was correct according to the manufacturer's recommendations. He soaked some utensils in the solution for 30 seconds. Were the utensils sanitized correctly?   Yes _____ No _____

② Josh mixed a chlorine sanitizer with 75°F (24°C) water. A test kit showed the concentration was 25 ppm. He soaked some tableware in the solution for 7 seconds. Was the tableware sanitized correctly?   Yes _____ No _____

③ Cecelia mixed an iodine sanitizer with 68°F (20°C) water. A test kit showed the concentration was 8 ppm. She put a pan in the solution for 30 seconds. Was the pan sanitized correctly?   Yes _____ No _____

④ Jarmin mixed a chlorine sanitizer with 100°F (38°C) water. A test kit showed the concentration was 50 ppm. She put a bowl in the solution for 7 seconds. Was the bowl sanitized correctly?   Yes _____ No _____

For answers, please turn to page 11.22.

Dishwashing

Tableware and utensils are often cleaned and sanitized in a dishwashing machine. Larger items such as pots and pans are often cleaned by hand in a three-compartment sink. Whichever method you use, you must follow specific practices so items are cleaned and sanitized. Then you must make sure you store the items so they do not become contaminated.

Machine Dishwashing

Dishwashing machines sanitize by using either hot water or a chemical sanitizing solution.

High-Temperature Machines

High-temperature machines use hot water to clean and sanitize. Therefore, water temperature is critical. If the water is not hot enough, items will not be sanitized. If the water is too hot, it may vaporize before the items have been sanitized. Extremely hot water can also bake food onto the items.

The temperature of the final sanitizing rinse must be at least 180°F (82°C). For stationary rack, single-temperature machines, it must be at least 165°F (74°C). The dishwasher must have a built-in thermometer, as shown in the photo at left, which checks water temperature at the manifold. This is where the water sprays into the tank. Operations that clean and sanitize a lot of tableware may need to install a heating device. This device will make sure the machines have enough hot water.

Chemical-Sanitizing Machines

Chemical-sanitizing machines can clean and sanitize items at much lower temperatures. Since sanitizers require different water temperatures, follow the dishwasher manufacturer's guidelines.

Dishwasher Operation

Operate your dishwasher according to the manufacturer's recommendations, and keep it in good repair. However, no matter what type of machine you use, you should follow these guidelines.

Keeping the machine clean   Clean the machine as often as needed, checking it at least once a day. Clear spray nozzles of food and foreign objects. Use a delimer to remove mineral deposits when needed. Fill tanks with clean water, and make sure detergent and sanitizer dispensers are filled.

Preparing items for cleaning   Scrape, rinse, or soak items before washing. Presoak items with dried-on food.

Loading dish racks   Use the right rack for the items you are washing, as shown in the photo at left. Load them so the water spray will reach all surfaces. NEVER overload dish racks.

Inspecting items   As each rack comes out of the machine, check it for dirty items. If you find some, run them through the machine again. Most items will need only one pass through the machine if the water temperature is right and the correct steps are followed.

Drying items   Air-dry all items. NEVER use a towel to dry items. You could recontaminate them.

Monitoring   Check water temperature and pressure. Follow the manufacturer's recommendations for dishwasher settings.

Manual Dishwashing

Operations often use a three-compartment sink to wash items. The sink station should have the parts shown in the illustration below.

- Area for rinsing away food or scraping it into garbage containers
- Drain board to hold dirty items
- Drain board to hold clean items
- Thermometer to measure water temperature
- Clock with a second hand to time how long items have been in the sanitizer

Before washing tableware, utensils, and equipment in a three-compartment sink, each sink and all work surfaces must be cleaned and sanitized.

How to Clean and Sanitize in a Three-Compartment Sink

Follow these steps to clean and sanitize items in a three-compartment sink.

1 Rinse, scrape, or soak items before washing them.

2 Clean items in the first sink. Wash them in a detergent solution at least 110°F (43°C). Use a brush, cloth, or nylon scrub pad to loosen dirt. Change the detergent solution when the suds are gone or the water is dirty.

3 Rinse items in the second sink. Spray them with water or dip them in it. Make sure you remove all traces of food and detergent. If dipping the items, change the rinse water when it becomes dirty or full of suds.

4 Sanitize items in the third sink following the guidelines on pages 11.4 and 11.5.

NEVER rinse items after sanitizing them. This could contaminate their surfaces. The only exception to this rule is when you are washing items in a dishwasher that can safely rinse items after they have been sanitized. The dishwasher must meet these requirements.

- The cycle sanitizes items before they are rinsed.
- The sanitizer is used according to the guidelines provided by the Environmental Protection Agency.

5 Air-dry items. Place items upside down so they will drain.

PATHOGEN PREVENTION

Storing Tableware and Equipment

Once utensils, tableware, and equipment have been cleaned and sanitized, they must be stored in a way that will protect them from contamination. Follow these guidelines.

Storage   Store tableware and utensils at least six inches (15 centimeters) off the floor. Protect them from dirt and moisture.

Storage surfaces   Clean and sanitize drawers and shelves before storing clean items.

Glasses and flatware   Store glasses and cups upside down on a clean and sanitized shelf or rack. Store flatware and utensils with handles up, as shown in the photo at left. Staff can then pick them up without touching food-contact surfaces, which will help prevent the transfer of pathogens such as Norovirus.

Trays and carts   Clean and sanitize trays and carts used to carry clean tableware and utensils. Check them daily, and clean as often as needed.

Stationary equipment   Keep the food-contact surfaces of stationary equipment covered until ready for use.

Apply Your Knowledge

The New Dishwasher

On a separate sheet of paper, list the missing or wrong steps in the story below.

Evan started work just as the breakfast rush had begun. A load of dirty dishes had just been put into the new dishwasher. There already were a lot of pots and pans to wash in the three-compartment sink, so Evan quickly got started. He scraped the dishes into a garbage container and stacked them on the drain board next to the first sink compartment. Then he filled the first compartment with hot water and added dish detergent. He put several pans in the soapy water to soak.

Next, Evan filled the remaining two compartments with warm water. He added iodine sanitizer to the third compartment. He used a thermometer to check the water temperature and then a test kit to check the sanitizer concentration. Both were good.

Using a nylon scrub pad, Evan worked on the pans until they were clean. As he finished each one, he dipped it in the sanitizing solution. Since customers had complained of an iodine flavor on tableware, he wanted to make sure there was no sanitizer left on the pans. As he pulled each pan out of the sanitizer, he placed it into the rinse water to soak for a few seconds. Then he put it on the clean drain board to air-dry.

About the time that Evan finished most of the pots and pans, the load in the dishwasher was done. Since the dirty dishes had started to pile up while Evan worked on the pans, he quickly unloaded the machine and got a dish cart. He noticed a few crumbs on the cart before he started. To clean it, he dipped a towel in the dishwater and wiped off the crumbs.

In the meantime, the carts of dirty dishes had grown. He quickly loaded a dishwasher rack with as many dishes as he could fit into it. Evan glanced into the dishwasher before pushing in the rack. He noticed some food bits stuck to a spray arm. He closed the door and started the load.

What did Evan do wrong?

What's Wrong with This Picture?

There are several things wrong with this three-compartment sink. Identify as many as you can in the space provided.

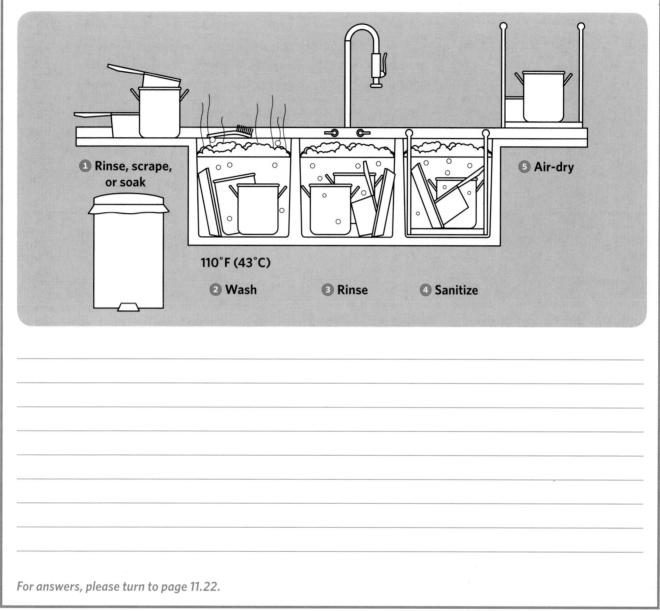

1 **Rinse, scrape, or soak**

110°F (43°C)

2 **Wash**   3 **Rinse**   4 **Sanitize**

5 **Air-dry**

For answers, please turn to page 11.22.

Cleaning and Sanitizing in the Operation

Keeping your operation clean means using the right tools and supplies. You also must store these items after using them to prevent contamination. Many of the chemicals you will use are hazardous, so you also have to know how to handle them to prevent injury.

For all your cleaning efforts to come together, you need a master cleaning schedule. Making this schedule work also means training and monitoring your employees to be sure they can follow it.

Cleaning the Premises

Nonfood-contact surfaces must be cleaned regularly. Examples include floors, ceilings, equipment exteriors, restrooms, and walls, as shown in the photo at left. Regular cleaning prevents dust, dirt, food residue, and other debris from building up.

Cleaning Tools and Supplies

Your staff needs many tools and supplies to keep the operation clean. They must also know how to use and store these items.

Choosing the Right Tools

Even a cleaning tool can contaminate surfaces if it is not handled carefully. Cleaning tools before putting them away can help prevent this. So does assigning tools for specific tasks. For example, you could use one set of tools to clean food-contact surfaces and another set for nonfood-contact surfaces. You could also use one set of tools for cleaning and another for sanitizing. Color-coding, as shown in the photo at left, often makes it easier for staff to know which set they should use. Whatever you do, always use a separate set of tools for the restroom.

Brushes   Brushes loosen dirt better than towels because they let you apply more pressure. However, worn brushes will not clean well and can contaminate surfaces. Throw them away. Always use the right brush for the job.

Scouring pads   Steel wool and other scouring pads are sometimes used to clean very dirty pots, pans, and equipment. However, metal scouring pads can break apart. If they are used on food-contact surfaces, bits of the pad can get into food. Nylon scouring pads are an alternative.

Mops and brooms   Keep both light- and heavy-duty mops and brooms on hand. Mop heads can be all cotton or synthetic blends. Have a different bucket, wringer, and mop for the front and back of the house. Both vertical and push-type brooms are also useful.

Towels   **Never** use towels meant for cleaning food spills for any other purpose. Store towels in a sanitizer solution between uses. Keep towels that come in contact with raw meat, fish, or poultry separate from other cleaning towels.

Storing Cleaning Tools and Supplies

Store cleaning tools and chemicals in a separate area away from food and prep areas. The storage area should have the following.

- Good lighting so employees can see chemicals easily
- Hooks for hanging mops, brooms, and other cleaning tools
- Utility sink for filling buckets and washing cleaning tools
- Floor drain for dumping dirty water, as shown in the photo at left

To prevent contamination, **never** clean mops, brushes, or other tools in sinks used for handwashing, food prep, or dishwashing. Additionally, **never** dump mop water or other liquid waste into toilets or urinals.

When storing cleaning tools, consider the following.

- Air-dry towels overnight.
- Hang mops, brooms, and brushes on hooks to air-dry.
- Clean and rinse buckets. Let them air-dry, and then store them with other tools.

Using Foodservice Chemicals

Many of the chemicals used in the operation can be hazardous, especially if they are used the wrong way. To reduce your risk, you should only use chemicals that are approved for use in a restaurant or foodservice operation. You should also follow these guidelines.

Storage and labeling   Store chemicals in their original containers away from food and prep areas, as shown in the photo at left. If chemicals are transferred to a new container, the label on that container must list the common name of the chemical.

Disposal   When throwing away chemicals, follow the instructions on the label and any requirements from your local regulatory authority.

Material Safety Data Sheets   The Occupational Safety and Health Administration (OSHA) has requirements for using chemicals. OSHA requires chemical manufacturers and suppliers to provide a Material Safety Data Sheet (MSDS) for each hazardous chemical they sell. An MSDS contains the following information about the chemical.

- Safe use and handling
- Physical, health, fire, and reactivity hazards
- Precautions
- Appropriate personal protective equipment (PPE) to wear when using the chemical
- First-aid information and steps to take in an emergency
- Manufacturer's name, address, and phone number
- Preparation date of MSDS
- Hazardous ingredients and identity information

MSDS are often sent with the chemical shipment. You also can request them from your supplier or the manufacturer. Employees have a right to see an MSDS for any hazardous chemical they work with. Therefore, you must keep these sheets where they can be accessed. The photo at left shows how one operation makes them available to employees.

Developing a Cleaning Program

To develop an effective cleaning program for your operation, you must focus on three things.

❶ Creating a master cleaning schedule

❷ Training your employees to follow it

❸ Monitoring the program to make sure it works

Creating a Master Cleaning Schedule

First, evaluate your cleaning needs. Walk through the facility and look at the way cleaning is done, as the manager in the photo at left is doing. Then figure out how things need to be cleaned and ways to improve cleaning processes. Next, make a master cleaning schedule. The schedule should have the following information.

What should be cleaned   Arrange the cleaning schedule in a logical way, so nothing is left out. List all cleaning jobs in one area. Or list jobs in the order they should be performed.

Who should clean it   Assign each task to a specific individual.

When it should be cleaned   Employees should clean and sanitize as needed. Schedule major cleaning when food will not be contaminated or service will not be affected. This is often after closing. Schedule work shifts to allow enough time.

How it should be cleaned   Have clear, written procedures for cleaning. Make sure they lead employees through the procedure step by step. List cleaning tools and chemicals by name. Post cleaning instructions near the item. Always follow manufacturers' instructions when cleaning equipment.

Training Your Employees to Follow the Program

Training is critical to the success of a cleaning program. Employees must know what you want them to do and how to do it. To help your program succeed, follow these guidelines.

Introducing the program   Schedule a kickoff meeting to introduce the program to employees. Explain the reason behind it. Stress how important cleanliness is to food safety. If people understand why they are supposed to do something, they are more likely to do it.

Training staff   Schedule time for training. Work with small groups or conduct training by area. Show staff how to clean in each area. The manager in the photo at left is training staff to clean a slicer.

Motivating staff   Provide a lot of motivation. Create incentives , such as a "Clean Team of the Month" award. Link performance to goals. For example, you might set a goal of high marks during inspections.

Monitoring the Cleaning Program

Once you have put the cleaning program in place, keep an eye on things to make sure it is working. You should check several areas.

- Supervise daily cleaning routines.
- Check all cleaning tasks against the master schedule every day.
- Change the master schedule as needed for any changes in menu, procedures, or equipment.
- Ask staff during meetings for input on the program.

Apply Your Knowledge

Is It the Right Tool?

Place a ✔ next to the situation if the employee used the right cleaning tool for the job.

① _____ Frederico used a metal scouring pad to clean pots and pans in the three-compartment sink. The pad left behind a few metal bits in the items, which he rinsed out.

② _____ Kathy used the same brush to scrub the kitchen walls every week for six months. The brush is worn, but she uses it anyway.

③ _____ Isabella has to mop the entrance, but she can't find the mop for the front of the house. Eager to go home, she mops the floor with the same mop she used in the kitchen.

Is It Stored Correctly?

Place a ✔ next to the situation if the employee stored the cleaning tool or material the right way.

① _____ Sheryl received a shipment of cleaning supplies. Along with the invoice, the supplier gave her an MSDS for the new brand of delimer she ordered. She filed the MSDS with the invoice in a locked cabinet.

② _____ Raul noticed that a bottle of delimer in the storage area was leaking. Fortunately, there was a nearly empty spray bottle of the same cleaner, so he poured the remainder into it. The label on the spray bottle listed the common name of the chemical.

③ _____ Sasha emptied a bucket of dirty mop water into the floor drain in the chemical-storage room. He rinsed the mop and hung it to dry. Then he cleaned and rinsed the bucket.

④ _____ Laura sanitized some of the food-contact surfaces on the shake machine using a spray bottle full of sanitizer. When she was finished, she placed the bottle on top of the machine so it would be handy the next time she needed it.

⑤ _____ Maurice used some delimer on the dishwasher. The sprayer on the bottle stopped working when it was only half empty, so he threw it in the garbage.

What's Wrong with This Picture?

There are many things wrong with this storage area. Identify as many as you can in the space provided.

_____

_____

_____

_____

_____

What Goes into a Master Cleaning Schedule?

Place a ✔ next to the minimum information needed in a master cleaning schedule.

① _____ Items to be cleaned

② _____ Steps for cleaning each item

③ _____ Date the schedule was created

④ _____ MSDS for each cleaning product

⑤ _____ When an item should be cleaned

⑥ _____ Tools and cleaning products for each item

⑦ _____ Staff member who should clean each item

⑧ _____ Emergency phone numbers in case of an accident

⑨ _____ Performance reviews of each staff member on the schedule

⑩ _____ Name and address of the manufacturer of each cleaning product

For answers, please turn to page 11.23.

Chapter Summary

Cleaning removes food and other dirt from a surface. Sanitizing reduces the number of harmful pathogens on a surface to safe levels. You must clean and rinse a surface before it can be sanitized. Then the surface must be allowed to air-dry. Surfaces can be sanitized with hot water or a chemical-sanitizing solution.

All surfaces should be cleaned and rinsed on a regular basis. Food-contact surfaces must be cleaned and sanitized after every use. You should also clean and sanitize each time you begin working with a different type of food or when a task is interrupted. If items are in constant use, they must be cleaned and sanitized every four hours.

Dishwashing machines can be used to clean, rinse, and sanitize most tableware and utensils. Always follow manufacturers' instructions. Make sure your machine is clean and in good working condition. Check the temperature and pressure of wash and rinse cycles daily.

Items cleaned in a three-compartment sink should be presoaked or scraped clean. They should then be washed in a detergent solution and rinsed in clean water. Finally, they should be sanitized in either hot water or in a chemical-sanitizing solution for a specific amount of time.

Cleaning tools and chemicals should be placed in a storage area away from food and food-prep areas. Make sure chemicals are clearly labeled. Keep MSDS for each chemical in a location accessible to all employees while on the job.

Develop and implement a cleaning program. Identify cleaning needs by walking through the operation and talking to employees. Create a master cleaning schedule listing all cleaning tasks. Explain to employees the important relationship between cleaning and sanitizing and food safety. Employees should be rewarded for good performance. Monitor the cleaning program to keep it effective, and supervise cleaning procedures. Make adjustments as needed.

Chapter Review Case Study

Now take what you have learned in this chapter and apply it to the following case study.

Tom was just hired as the new general manager at the Twin Trees Family Restaurant. One of his first projects was to create a new cleaning program. He started by taking a walk through the operation. His first stop was the storage area for cleaning tools and supplies. It had a utility sink and a floor drain, but the hot water in the sink wasn't working. He also noticed two sets of mops and brooms stored on the floor. One set was labeled "Restroom Only." The storage area was small, but it was well organized and well lit. All the containers were clearly labeled.

① Should Tom suggest any changes to the storage room, tools, or chemicals?   Yes _____   No _____
If yes, what changes should he suggest?

_____

_____

_____

_____

_____

Next, Tom watched how Valerie, a buser, cleaned and sanitized her areas. In the dining room, Valerie needed to clean up after a family with young children. After clearing the table, she wiped it with a rag stored in a detergent solution. After this, she carried a tray of dirty dishes back to the dishwashing area. The staging area for the dishwasher was full, so she placed the tray on the drain board by the three-compartment sink. Then she went to the storage area, where she filled a bucket with cold water. Unable to find the general-purpose cleaner she normally uses to clean floors, she took a box of heavy-duty cleaner and put some in the bucket. Then she got a mop and broom and headed back to the dining room. She swept up the crumbs and mopped the floor around the table. After this, she brought the tools back to the storage area, dumped the mop water in the utility drain, and hung up the broom. Wanting to rinse out the mop in hot water, she took the bucket to the dishwashing sink, where she filled it with very hot water. After she rinsed the mop, she dumped the water in the drain, wrung out the mop, and left it in the wringer to dry.

② Did Valerie do anything wrong?   Yes _____   No _____   If yes, what should she have done?

_____

_____

_____

_____

Next, Tom watched Clara, a new prep cook, to see how she cleaned and sanitized her areas. Clara cut some melons on a cutting board. Then she wiped it down with a towel. Clara put the towel in a bucket of sanitizing solution to soak while she butterflied some pork chops on the board. Using the same towel, she wiped down the board after she finished the pork chops. Then, she chopped some onions and sautéed them in a large stockpot. While the onions were sautéing, Clara wiped the board a third time with the same towel.

③ Did Clara do anything wrong?   Yes _____   No _____   If yes, what should she have done?

_____

_____

_____

_____

Tom also watched many other staff members perform cleaning and sanitizing tasks that week. With the help of some senior staff, Tom created a master cleaning schedule.

④ What steps should Tom take to make sure everyone follows the master cleaning schedule?

_____

_____

_____

_____

_____

For answers, please turn to page 11.23.

Study Questions

Circle the best answer to each question below.

① **What is sanitizing?**

A Reducing dirt from a surface

B Reducing the pH of a surface

C Reducing the hardness of water

D Reducing pathogens to safe levels

② **If food-contact surfaces are in constant use, how often must they be cleaned and sanitized?**

A Every 4 hours

B Every 5 hours

C Every 6 hours

D Every 7 hours

③ **An employee wants to make a sanitizer solution to spray onto food-contact surfaces. What must be done to ensure that it has been made correctly?**

A Test the solution with a sanitizer test kit.

B Use very hot water when making the solution.

C Try out the solution on a food-contact surface.

D Compare the color to another solution of the right strength.

④ **How can foodservice managers find out which chemical sanitizers are appropriate for their operations?**

A Check with the Food and Drug Administration.

B Check with the Centers for Disease Control and Prevention.

C Check with the local regulatory authority.

D Check the label on the sanitizer container.

⑤ **What should be done when throwing away chemicals?**

A Pour leftover chemicals into a drain and throw the container away.

B Seal the container in a bag and put it next to the garbage container.

C Follow label instructions and any regulatory requirements that apply.

D Take the lid off the container and put it into a garbage container.

⑥ **Material Safety Data Sheets (MSDS) should be**

A kept so employees can access them.

B kept with paperwork from the supplier.

C sent to the local regulatory authority.

D memorized in case of an emergency.

⑦ **Flatware and utensils that have been cleaned and sanitized should be stored**

    A with the handles facing up.

    B at a temperature of 41°F (5°C) or lower.

    C above cleaning supplies.

    D within six inches (15 centimeters) of the floor.

⑧ **What is the correct way to clean and sanitize a prep table?**

    A Wash, rinse, sanitize, air-dry

    B Rinse, wash, sanitize, air-dry

    C Sanitize, wash, rinse, air-dry

    D Air-dry, rinse, sanitize, wash

For answers, please turn to page 11.24.

Answers

11.6   Take the Right Steps

3, 1, 4, 2

11.6   To Sanitize or Not to Sanitize

2, 3, and 4 should be marked.

11.6   Which Cleaner Should You Use?

① D

② B

③ C

④ A

11.6   Was It Sanitized?

① Yes

② No

③ No

④ Yes

11.10   The New Dishwasher

Here is what Evan did wrong.

- He did not clean and sanitize the sink compartments and drain boards before starting.

- He did not check the water temperature in the first sink compartment.

- He did not rinse the items before sanitizing them. He rinsed the items after sanitizing, which could contaminate them.

- He did not time how long the pots and pans were in the sanitizer.

- He did not clean and sanitize the cart for clean tableware.

- He did not rinse, scrape, or soak the dirty dishes before putting them into the dish rack.

- He overloaded the dish rack.

- He did not clean off the dishwasher spray arm before starting the load.

11.11   What's Wrong with This Picture?

① There is no clock with a second hand. Employees would not be able to time how long an item has been immersed in the sanitizer.

② Soap suds from the wash compartment have been carried over into the rinse compartment and the sanitizer compartment. This can deplete the sanitizer.

③ A cleaned and sanitized pot is not being air-dried properly. It should be inverted.

11.16   Is It the Right Tool?

None of the employees used the right tools.

11.16   Is It Stored Correctly?

2 and 3 should be marked.

11.17   What's Wrong with This Picture?

① There are no hooks for the brushes and mop to air-dry.

② The chemical spray bottle is not labeled.

③ Food is being stored in the area.

④ There should be separate tools for cleaning the restrooms, the back of the house, and the front of the house.

11.17   What Goes into a Master Cleaning Schedule?

1, 2, 5, 6, and 7 should be marked.

11.18   Chapter Review Case Study

① Yes. Tom should make the following suggestions.

- Fix the hot water.
- Have a separate set of tools for the front and the back of the house (in addition to the set for the restroom).
- Install hooks for the mops and brooms.

② Yes. Valerie should have done the following.

- She should have checked if the heavy-duty cleaner was safe to use in place of the all-purpose cleaner.
- She should have stored the rag in a sanitizing solution.
- She should have cleaned the bucket before storing it.
- She should not have filled the bucket in the dishwashing sink.
- She should have hung the mop to dry.

③ Yes. Clara should have washed, rinsed, and sanitized the cutting board at these times.

- Before cutting the melons
- After cutting melons and before butterflying pork chops
- After butterflying pork chops and before chopping onions
- After chopping onions.

Continued on the next page ▶

▶ *Continued from previous page*

④ For Tom's cleaning program to work, he should do the following.

- Have a meeting to introduce the program.
- Train the staff on the cleaning and sanitizing tasks.
- Motivate the staff to follow the program.
- Monitor the program.

11.20   Study Questions

① D

② A

③ A

④ C

⑤ C

⑥ A

⑦ A

⑧ A

Notes

12

Integrated Pest Management

In the News

Restaurant Back in Business After Rodent Troubles

A local restaurant reopened after closing due to health-code violations. The health inspector, who had found no evidence of rodents in an inspection six months earlier, reported a widespread mouse problem. There were mouse feces on food boxes, clean dishes, shelves, trays, prep tables, and along walls. The inspector also saw a live mouse in an unlit storage area.

After the inspection, the restaurant's manager agreed to close the restaurant for cleaning and pest-control measures. A follow-up inspection took place before the restaurant reopened.

You Can Prevent This

The restaurant in the story above could have prevented a health-code violation by denying pests entry to the facility or by having better pest-control procedures in place. To prevent a pest infestation in your operation, you need to learn about the following topics.

- Implementing an integrated pest management (IPM) program

- Working with a pest control operator (PCO)

Concepts from Earlier Chapters

Before reading this chapter, remember these concepts and facts.

Cleaning and sanitizing   Cleaning removes food and dirt from a surface. Sanitizing reduces pathogens on a surface to safe levels.

FIFO   Method of rotating refrigerated, frozen, and dry food during storage so the oldest inventory is used first.

Material Safety Data Sheets   MSDS identify the hazards of using a chemical and give directions for safe use and handling.

The Integrated Pest Management (IPM) Program

Rodents, insects, birds, and other pests are more than just unsightly to customers. They can damage food, supplies, and facilities. But the greatest danger comes from their ability to spread diseases, including foodborne illnesses. The best way to deal with pests is to have an IPM program.

Rules of an IPM Program

An IPM program has two parts. First, it uses prevention measures to keep pests from entering the operation. Second, it uses control measures to eliminate any pests that do manage to get inside.

Prevention is critical in pest control. Don't wait until you find pests in your operation. If you do see them, they may already be present in large numbers. Once this happens you have an infestation, and an infestation can be very difficult to eliminate.

Control measures are needed for any pests that do manage to get into your operation. For your IPM program to be successful, you should work with a licensed pest control operator (PCO), as shown in the photo at left. These professionals use safe methods to prevent and control pests.

There are three basic rules for an IPM program.

❶ Deny pests access to the operation.

❷ Deny pests food, water, and a hiding or nesting place.

❸ Work with a licensed PCO to eliminate pests that do enter the operation.

Keeping Pests Out of the Operation

Pests can enter an operation in one of two ways. Sometimes they are brought inside with deliveries. They can also enter through openings in the building. Prevent pests from entering by paying attention to the following areas.

Deliveries

- Use approved, reputable suppliers.

- Check all deliveries before they enter your operation, as shown in the photo at left.

- Refuse shipments in which you find pests or signs of pests. This includes egg cases and body parts (legs, wings, etc.).

PATHOGEN PREVENTION

Doors, Windows, and Vents

- Screen all windows and vents with at least 16 mesh per square inch screening. Larger mesh sizes can let in mosquitoes or flies, which can lead to contamination from bacteria, such as *Shigella* spp. Check screens regularly, and clean, patch, or replace them as needed.

- Install self-closing devices and door sweeps on all doors. Repair gaps and cracks in doorframes and thresholds. Use weather stripping on the bottom of doors with no thresholds.

- Install air curtains (also called air doors or fly fans) above or alongside doors. These devices blow a steady stream of air across the entryway. This creates an air shield around doors that stops insects from entering.

- Keep all exterior openings closed tightly. Drive-thru windows should be closed when not in use, as shown in the photo at left.

Pipes

- Mice, rats, and insects use pipes as highways through a facility.

- Use concrete to fill holes or sheet metal to cover openings around pipes. The photo at left shows an example.

- Install screens over ventilation pipes and ducts on the roof.

- Cover floor drains with hinged grates to keep rodents out. Rats are very good swimmers and can enter buildings through drainpipes.

Floors and Walls

- Rodents can burrow into buildings. They can dig through decaying masonry or cracks in building foundations. They also move through floors and walls in the same way. Mice can squeeze through holes the size of a nickel. Rats can pass through holes the size of a half dollar.

- Seal all cracks in floors and walls. Use a permanent sealant recommended by your PCO or local regulatory authority.

- Seal spaces or cracks where stationary equipment is fitted to the floor. Use an approved sealant or concrete, depending on the size of the gaps.

Denying Pests Food and Shelter

Pests are usually attracted to damp, dark, and dirty places. A clean operation offers them little access to food and shelter. The stray pest that might get in cannot survive or breed in a clean kitchen. Besides sticking to your master cleaning schedule, follow these guidelines.

Garbage disposal   Throw out garbage quickly and correctly. Garbage attracts pests and provides them with a place to breed. The photo at left shows you what NOT to do. Keep garbage containers clean and in good condition. Keep outdoor containers tightly covered. Clean up spills around garbage containers immediately, and wash containers regularly.

Recyclables   Store recyclables in clean, pest-proof containers. Keep them as far away from your building as local regulations allow. Bottles, cans, paper, and packaging give pests food and shelter.

Food and supplies   Store all food and supplies the right way and as quickly as possible.

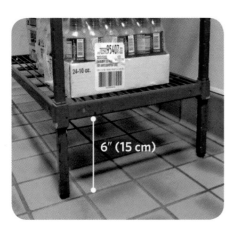

6" (15 cm)

* Keep food and supplies away from walls and at least six inches (15 centimeters) off the floor, as shown in the photo at left.

* When possible, use dehumidifiers to keep humidity at 50 percent or lower. Low humidity helps prevent roach eggs from hatching.

* Consider refrigerating food such as powdered milk, cocoa, and nuts after opening. Most insects that might be attracted to this food become inactive at temperatures below 41°F (5°C).

* Use FIFO to rotate products so pests do not have time to settle into them and breed.

Cleaning   Careful cleaning eliminates the pests' food supply and destroys insect eggs. It also reduces the places pests can take shelter.

* Clean up food and beverage spills immediately, including crumbs and scraps.

* Clean toilets and restrooms as often as necessary.

* Train employees to keep lockers and break areas clean. Food and dirty clothes should not be kept in or around lockers. Remove garbage and food debris from break rooms regularly.

* Keep cleaning tools and supplies clean and dry. Store wet mops on hooks rather than on the floor, since roaches can hide in them.

* Empty water from buckets to keep from attracting rodents.

Apply Your Knowledge

The Rules of Integrated Pest Management

There are 13 pest-related situations listed below. For each situation, place a ✔ next to the IPM rule that should be applied to the situation.

① An employee notices a hole in the foundation of the operation.

_____ Deny pests access.   _____ Deny pests food/water/hiding place.   _____ Work with a PCO to eliminate pests.

② A delivery of tomatoes arrives and has fruit flies buzzing around it.

_____ Deny pests access.   _____ Deny pests food/water/hiding place.   _____ Work with a PCO to eliminate pests.

③ Some of the staff have propped the back door open because it is hot inside.

_____ Deny pests access.   _____ Deny pests food/water/hiding place.   _____ Work with a PCO to eliminate pests.

④ A mop is stored with the wet mop head on the floor of the cleaning-supplies closet.

_____ Deny pests access.   _____ Deny pests food/water/hiding place.   _____ Work with a PCO to eliminate pests.

⑤ A floor drain is missing its cover.

_____ Deny pests access.   _____ Deny pests food/water/hiding place.   _____ Work with a PCO to eliminate pests.

⑥ A window at the back of the operation is missing a screen.

_____ Deny pests access.   _____ Deny pests food/water/hiding place.   _____ Work with a PCO to eliminate pests.

⑦ There are full garbage bags piled up by the back door.

_____ Deny pests access.   _____ Deny pests food/water/hiding place.   _____ Work with a PCO to eliminate pests.

⑧ A pipe entering the operation has a large gap all the way around it.

_____ Deny pests access.   _____ Deny pests food/water/hiding place.   _____ Work with a PCO to eliminate pests.

⑨ An employee sees a cockroach in the kitchen.

_____ Deny pests access.   _____ Deny pests food/water/hiding place.   _____ Work with a PCO to eliminate pests.

⑩ A bag of recyclables has been left leaning against the back of the building.

_____ Deny pests access.   _____ Deny pests food/water/hiding place.   _____ Work with a PCO to eliminate pests.

⑪ A delivery of produce arrives at the back of the loading dock. The receiving staff notices insect parts in some of the boxes.

_____ Deny pests access.   _____ Deny pests food/water/hiding place.   _____ Work with a PCO to eliminate pests.

⑫ A case of noodles has been stored on the floor of the dry-storage room.

_____ Deny pests access.   _____ Deny pests food/water/hiding place.   _____ Work with a PCO to eliminate pests.

⑬ A staff member sees a rat in the storeroom.

_____ Deny pests access.   _____ Deny pests food/water/hiding place.   _____ Work with a PCO to eliminate pests.

For answers, please turn to page 12.12.

Working with a Pest Control Operator (PCO)

Even after you have made every effort to keep pests out, they may still get into your operation. If this happens, you must work with a PCO to control pests, and you will need to know how to select a qualified one. You should also know how to determine what kind of pests are present. Finally, understanding the basics of using and storing pesticides is important for keeping both food and people safe.

Hiring a Pest Control Operator

Although you can take many preventive measures to reduce the risk of an infestation, most control measures must be carried out by a licensed PCO. A PCO will help you develop an integrated approach to pest management. This approach will use a combination of chemical and nonchemical treatments.

Hiring a PCO is like choosing any other service provider. You must do your homework. Check references and make sure that the PCO is licensed, if required by your state. You also might want a PCO who is a member of a pest management association, or who is certified by one. Before signing a contract, make sure you understand what the pest problem is, how bad the problem is, and what must be done to get rid of it.

Identifying Pests

To work with your PCO effectively, you must know how to determine the type of pests you are dealing with. Record the time, date, and location of any signs of pests and report them to your PCO. Early detection means early treatment.

Cockroaches

Roaches often carry pathogens such as *Salmonella* spp., fungi, parasite eggs, and viruses. Most live and breed in dark, warm, moist, and hard-to-clean places. If you see a cockroach in daylight, you may have a major infestation. Generally, only the weakest roaches come out during the day. If you think you have a roach problem, check for the following signs, as shown in the photo at left.

- Strong, oily odor

- Droppings (feces) that look like grains of black pepper

- Capsule-shaped egg cases that are brown, dark red, or black and may appear leathery, smooth, or shiny

Rodents

Rodents are a serious health hazard. They eat and ruin food, damage property, and spread disease. A building can be infested with both rats and mice at the same time. Look for the following signs.

Gnaw marks   Rats and mice gnaw to get at food and to wear down their teeth, which grow continuously.

Droppings and urine stains   Fresh droppings are shiny and black. Older droppings are gray. Rodent urine will "glow" when exposed to a black (ultraviolet) light.

Tracks   Rodents tend to use the same pathways through your operation. If rodents are a problem in your operation, you may see dirt tracks along light-colored walls.

Nests   Rats and mice use soft materials, such as scraps of paper, cloth, hair, feathers, and grass, to build their nests. The photo at left shows an example of a mouse's nest.

Holes   Rats usually nest in holes located in quiet places. Nests are often found near food and water and may be found next to buildings.

Using Pesticides

Sometimes it may seem more cost effective to purchase and apply pesticides yourself. However, there are many reasons NOT to do this.

- Pesticides that are applied the wrong way may be ineffective or harmful.
- Pests can develop resistance and immunity to pesticides.
- Each region has its own pest-control problems, and some control measures are more effective than others.
- Pesticides are regulated by federal, state, and local laws. Some pesticides are not approved for use in foodservice operations.

Rely on your PCO to decide how pesticides should be used in your operation. They are trained to determine the best pesticide for each pest, and how and where to apply it. Have your PCO apply pesticides when you are closed for business and employees are not on-site. Follow these guidelines whenever pesticides are applied.

- Prepare the area to be sprayed by removing all food and movable food-contact surfaces.
- Cover equipment and food-contact surfaces that cannot be moved, as shown in the photo at left.
- Wash, rinse, and sanitize food-contact surfaces after the area has been sprayed.

Storing Pesticides

Your PCO should store and throw out all pesticides used in your facility. If pesticides are stored on the premises, use the following guidelines.

• Keep pesticides in their original containers.

• Store pesticides in a secure location away from areas where food, utensils, and food equipment are stored.

• Check local regulations before throwing out pesticides. Many chemicals are considered hazardous waste. Throw out empty containers according to manufacturers' directions and local regulations.

• Keep a copy of the pesticides' Material Safety Data Sheets (MSDS) on the premises.

Apply Your Knowledge

Who Am I?

Write a **C** next to the statement if it applies to cockroaches. Write an **R** next to the statement if it applies to rodents.

① _____ I nest in scraps of paper, cloth, and hair.

② _____ I produce a strong, oily odor.

③ _____ I like to gnaw on things.

④ _____ My droppings are shiny black or grey.

⑤ _____ My nests are often found next to buildings.

⑥ _____ My droppings look like grains of pepper.

⑦ _____ I live in dark, warm, moist places.

⑧ _____ If you see me in daytime, I am present in large numbers.

Doing It the Right Way

Place a ✔ next to the right things to do to control pests.

① _____ A shift manager applies pesticides after closing for the evening.

② _____ A PCO applies pesticides after the operation has closed for the evening.

③ _____ Pesticides are applied to the kitchen during operating hours.

④ _____ Food-contact surfaces are washed, rinsed, and sanitized after pesticides have been applied.

⑤ _____ Pesticides are stored in their original containers.

⑥ _____ Pesticides are stored under food, on the bottom shelf.

⑦ _____ Pesticides are stored in a secure location, away from food.

⑧ _____ A manager keeps a copy of the pesticide MSDS on-site.

⑨ _____ A PCO throws out leftover pesticide.

⑩ _____ A cook keeps a can of roach spray next to the grill.

For answers, please turn to page 12.12.

Chapter Summary

Pests can carry and spread many diseases. Once they have infested an operation, they can be very difficult to eliminate. You need to develop and put in place an integrated pest management (IPM) program. An IPM program uses prevention measures to keep pests from entering the operation. It also uses control measures to get rid of any pests that do get inside the operation.

To keep pests out, you must deny them food, water, and shelter. Refuse any shipment that has pests or signs of pests. Screen all windows and vents. Install self-closing doors and air curtains. Keep exterior openings closed when not in use. Fill or cover holes around pipes. Seal cracks in floors and walls. You should also work with a pest control operator (PCO) to eliminate any pests that have gotten into the operation.

Pests are usually attracted to damp, dark, and dirty places. So stick to your master cleaning schedule. Throw out garbage quickly. Keep outdoor containers clean and tightly covered. Keep food and supplies away from walls and at least six inches (15 centimeters) off the floor.

Rely on your PCO to decide how pesticides should be used in your operation. PCOs are trained to determine the best pesticide for each pest. They also know how and where to apply it. Your PCO should store and throw out all pesticides used in your facility. If they are stored on-site, they should be stored in a secure location away from areas where food, utensils, and food equipment are stored. Check local regulations before throwing out pesticides. Also keep a copy of the pesticides' MSDS on-site.

Chapter Review Case Study

Now take what you have learned in this chapter and apply it to the following case study.

Jorge had been chatting with the managers of the businesses near his operation, and he didn't like what he had heard. "Rats," they all said. Both the convenience store on one side of his operation and the quick-service place on the other side had a rat problem. Jorge knew that he hadn't really been thinking about pests recently. "Maybe I need to put together an IPM program," he thought. It was time to inspect his operation and see what he could do to prevent a pest problem.

Jorge started his assessment outside. As he stepped out of the operation, the produce supplier arrived with the daily delivery. Jorge waved the driver in and asked him to stack everything in the back of the building. He'd inspect the produce after it was safely inside. At the back of the operation, Jorge noted that a couple of garbage bags were lined up neatly at the back door. The kitchen staff would take them to the outside bins when the pile got too big. This was more convenient than carrying them all the way outside each time. "Well, that looks pretty good," he thought. Then Jorge pushed some of the bags aside and saw a large crack in the foundation. He made a mental note to fill in the crack to keep rodents out.

Walking through the back of the house, Jorge paused to look under the equipment. He did not like what he saw. The employee who cleaned up at night was definitely not sweeping and mopping underneath the equipment. There was a thick layer of food, grime, and even some silverware under the dishwasher. It looked like it had been there for awhile. "How many times do I have to tell this guy to mop UNDER the equipment?" he asked himself. Shaking his head, Jorge moved on to the storeroom.

Jorge realized that it had been awhile since they had rearranged the storeroom. In the corner was an old stack of plastic milk crates that had been there for at least a year. He pulled the crates away from the wall. There he saw a

Continued on the next page ▶

► *Continued from previous page*

rat's nest. Looking closer he also found specks that looked like black pepper along the wall, where the wall met the floor. "Great," he muttered. "Rats AND roaches."

He knew what he had to do. The pesticides were on the bottom shelf of the storeroom, right underneath the hamburger buns. Jorge shrugged his shoulders and said, "Looks like I'm going to spend tomorrow spraying the place down with pesticides. Maybe I'll do it after breakfast, before the lunch rush."

① What did Jorge do right?

_____

_____

_____

_____

_____

② What should Jorge have done differently?

_____

_____

_____

_____

_____

For answers, please turn to page 12.13.

Study Questions

Circle the best answer to each question below.

① **Who should apply pesticides?**

A Shift manager

B Person in charge

C Pest control operator

D Designated pest employee

② **Cockroaches typically are found in places that are**

A cold, dry, and light.

B warm, dry, and light.

C cold, moist, and dark.

D warm, moist, and dark.

③ **What kind of odor is a sign that roaches might be present?**

A Strong, oily

B Warm, spicy

C Sharp, musty

D Mild, seaweed

④ **The three basic rules of an integrated pest management program are 1) deny pests access to the operation, 2) _____ and 3) work with a licensed PCO to eliminate pests that do enter.**

A deny pests food, water, and a nesting or hiding place

B document all infestations with the local regulatory authority

C prepare a chemical-application schedule and post it publicly

D notify the EPA that pesticides are being used in the establishment

⑤ **After pesticides have been applied, food-contact surfaces should be**

A used only after a 20-minute wait.

B checked with a sanitizer test kit.

C washed, rinsed, and sanitized.

D replaced with new equipment.

⑥ **If pesticides are stored in the operation, where should they be kept?**

A In a secure location, away from food

B In a glass container, in a walk-in cooler

C In dry storage, on a shelf below the food

D In a plastic container, in any location

For answers, please turn to page 12.13.

Answers

12.5   **The Rules of Integrated Pest Management**

   ① Deny pests access.

   ② Deny pests access.

   ③ Deny pests access.

   ④ Deny pests food/water/hiding place.

   ⑤ Deny pests access.

   ⑥ Deny pests access.

   ⑦ Deny pests food/water/hiding place.

   ⑧ Deny pests access.

   ⑨ Work with a PCO to eliminate pests.

   ⑩ Deny pests food/water/hiding place.

   ⑪ Deny pests access.

   ⑫ Deny pests food/water/hiding place.

   ⑬ Work with a PCO to eliminate pests.

12.8   **Who Am I?**

   ① R

   ② C

   ③ R

   ④ R

   ⑤ R

   ⑥ C

   ⑦ C

   ⑧ C

12.8   **Doing It the Right Way**

   2, 4, 5, 7, 8, and 9 should be marked.

12.9   Chapter Review Case Study

① Here is what Jorge did right.

- Jorge recognized the need to develop an IPM program for his operation.

- While inspecting his building, he recognized that the crack in the foundation was a possible entry point for pests and should be filled in.

- Jorge also inspected his kitchen carefully and recognized the grime under the dishwasher as a hazard that could provide food for pests. He clearly understands the need to deny access, food, and shelter to pests.

- When inspecting his storeroom, Jorge recognized the signs of rodents and cockroaches.

② Here is what Jorge should have done differently.

- While Jorge recognized the need to develop an IPM program, he should have had one in place already. When the produce supplier arrived, Jorge allowed the supplier to bring produce into the operation without first inspecting it. This provided a potential opportunity for pests to enter the operation with the delivery.

- Garbage bags were stacked by the back door. This can provide food and shelter for pests. Garbage should always be removed quickly from the operation and stored in tightly covered, outdoor containers.

- The old milk crates in the storeroom, plus the rat and cockroach feces, indicate that area had not been cleaned recently. Jorge should have made sure that all areas of the storeroom were cleaned regularly. The old crates should have been immediately returned to the supplier or thrown out to deny pests shelter.

- Jorge stored pesticides near food. They should have been stored in a secured location, away from where food is stored and prepped.

- Jorge planned to apply pesticides himself. He should have contacted a PCO to do it, to make sure that the right pesticides are applied safely.

- Jorge was planning to apply pesticides during operating hours. This should be done when the kitchen is not in operation, once equipment and food-contact surfaces have been covered and customers and staff are not present.

12.11   Study Questions

① C   ④ A

② D   ⑤ C

③ A   ⑥ A

IV Food Safety Regulations and Employee Training

13

Food Safety Regulations and Standards

Working Together to Solve the Problem

A local Italian restaurant was cited by city health officials for cooling its meat sauce incorrectly. The sauce is used in most of the restaurant's dishes. To have enough of the sauce for the weekend rush, the staff typically made a large batch on Sunday night. They then stored it in five-gallon buckets until the next weekend.

The health inspector found that the sauce was still 70°F (21°C) at its center five days after being placed in the refrigerator. This could have led to a foodborne-illness outbreak, the inspector said.

Working with the health inspector, the owner solved the problem by adding ice to the sauce to help it cool and making smaller batches of sauce twice a week. This allowed the sauce to cool quickly enough to be safe.

You Can Build on This

The actions taken by the manager and the inspector in the story above may have prevented a foodborne-illness outbreak. As a manager, you must interact with health inspectors and the regulatory community. You will benefit from a good working relationship with them and from understanding their expectations. To do this, there are several things you should know.

- Who creates and enforces food safety regulations

- How regulatory and self-inspections work

Concepts from Earlier Chapters

Before reading this chapter, remember these concepts and facts.

CDC risk factors   Five most common factors responsible for foodborne illness, as identified by the Centers for Disease Control and Prevention (CDC).

Government Regulation of Foodservice Operations

A big part of your job is keeping food safe. So you must understand the rules and regulations that impact food safety. It is also helpful to know who makes these rules and who enforces them.

The *FDA Food Code*

The *FDA Food Code*, as shown in the photo at left, is issued by the Food and Drug Administration (FDA). It is based on input from the Conference for Food Protection (CFP). CFP representatives come from the food industry, government, academia, and consumer groups. The *FDA Food Code* outlines the federal government's recommendations for food safety regulations for the foodservice industry. It is not an actual law. Although the FDA recommends adoption by the states, it cannot require it.

The *FDA Food Code* has food safety guidelines for every part of a foodservice operation.

FDA Food Code Guidelines

Food Safety Topics	Areas Addressed
Foodhandling and preparation	Criteria for receiving, storage, display, service, and transportation
Personnel	Health, personal cleanliness, clothing, and hygiene practices
Equipment and utensils	Materials, design, installation, and storage
Cleaning and sanitizing	Facilities and equipment
Utilities and services	Water, sewage, plumbing, restrooms, waste disposal, and integrated pest management
Construction and maintenance	Floors, walls, ceilings, lighting, ventilation, dressing rooms, locker areas, and storage areas
Foodservice units	Both mobile and temporary units
Compliance procedures	Foodservice inspections and enforcement actions

State and Local Regulations

In the United States, most food regulations affecting foodservice operations are written at the state level.

- Each state decides whether to adopt the *FDA Food Code* or some modified form of it.

- State regulations may be enforced by state or local (city or county) regulatory authorities.

- Health inspectors from city, county, or state health departments conduct foodservice inspections in most states, as shown in the photo at left. Generally, they are trained in food safety, sanitation, and public health principles.

Apply Your Knowledge

Who's in Charge Here?

Match the correct answer to each question. Some answers may be used more than once.

Ⓐ Food and Drug Administration (FDA)

Ⓑ *FDA Food Code*

Ⓒ Foodservice manager

Ⓓ Health inspector

Ⓔ State government

① _____ A recommendation, not a law, that includes the government's food safety recommendations

② _____ Responsible for creating each operation's food safety practices

③ _____ Responsible for conducting foodservice inspections for the regulatory authority

④ _____ Decides whether or not to adopt the *FDA Food Code*

⑤ _____ Usually trained in food safety, sanitation, and public health principles

For answers, please turn to page 13.10.

Inspections

Inspections are a reality for every manager. Preparing for and understanding the inspection process are important tasks. Failing an inspection may result in the closure of your operation. You must know what can cause a closure. Equally important is understanding the role of self-inspection.

Why Regulatory Inspections Are Important

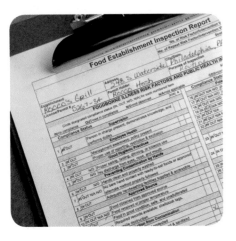

The most important reason for regulatory inspections is that failure to ensure food safety can risk the health of your customers. A lack of food safety could even cost you your business. Basically, an inspection evaluates whether an operation is meeting minimum food safety standards. It also produces a written report that notes deficiencies, as shown in the photo at left. The report will help you bring your operation into compliance with safe food practices.

The Inspection Process

All operations serving food to the public will receive an inspection. This includes everything from quick-service and fine-dining restaurants to delicatessens, hospitals, nursing homes, and schools.

Regulatory authorities have begun taking a more risk-based approach when conducting inspections. Inspections now address more than just normal compliance issues. Many regulatory authorities look at how the operation is managing risks, using the five Centers for Disease Control and Prevention (CDC) risk factors (which you learned in chapter 1) and the FDA's public health interventions (which you learned in chapter 9) as a guide.

The FDA recommends that regulatory authorities use the following three risk designations when evaluating establishments. These replace the "critical" and "noncritical" risk designations recommended previously.

- Priority items

- Priority foundation items

- Core items

Priority items are the most critical. These are actions and procedures that prevent, eliminate, or reduce hazards associated with foodborne illness to an acceptable level. Proper handwashing would be considered a priority item. Priority foundation items are those that support a priority item. Having soap at a handwashing sink is an example. Core items relate to general sanitation, the facility, equipment design, and general maintenance. Keeping equipment in good repair is an example.

Guidelines in the Inspection Process

In most cases, inspectors will arrive without warning. They will usually ask for the manager of the operation. Make sure your employees know who is in charge of food safety in your absence. Also, keep in mind any policies your company has on how to handle an inspection. The following guidelines can help you get the most out of food safety inspections.

Asking for identification   Don't let anyone enter the back of the operation without the right identification. Many inspectors will volunteer their credentials, as shown in the photo at left. Also, make sure you know the reason for the inspection. The visit may be a routine inspection or the result of a customer complaint, or it could be for some other purpose. Do NOT refuse entry to an inspector. In some areas, inspectors may have authority to gain access to the operation. They also may have authority to revoke the operation's permit for refusing entry.

Cooperating with the inspector   Answer all the inspector's questions to the best of your ability. Tell your employees to do the same. Go with the inspector during the inspection. You will be able to answer any questions and possibly correct problems immediately. If something can't be fixed right away, tell the inspector when it can be corrected. Open communication is important for building a good working relationship with the inspector. You'll also have the chance to learn from the inspector's comments and get food safety advice.

Taking notes   As you walk with the inspector, make note of any problems pointed out, as shown in the photo at left. This will help you remember exactly what was said. Make it clear that you are willing to fix any problems. If you believe the inspector is incorrect about something, note what was mentioned. Then contact the regulatory authority.

Keeping the relationship professional   Be polite and friendly, and treat inspectors with respect. Be careful about offering food, drink, or anything else that could be misunderstood as trying to influence the inspection report.

Being prepared to provide records requested by the inspector
An inspector might ask for many types of records.

- Purchasing records to make sure that food has been received from an approved source

- Pest control treatments

- List of chemicals used in the operation

- Proof of food safety knowledge, such as a food protection manager certificate, as shown in the photo at left

- HACCP records, in some cases

You might ask the inspector why these records are needed. If a request seems inappropriate, check with the inspector's supervisor.

You can also check with your lawyer about limits on confidential information. Remember, any records you give to the inspector will become part of the public record.

Discussing violations and time frames for correction with inspectors

After the inspection, the inspector will discuss the results and the score, if one is given. Study the inspection report closely. You must understand the exact nature of a violation. You must also know how a violation affects food safety, how to correct it, and whether or not the inspector will follow up.

You will be asked to sign the inspection report. Signing the report means acknowledging that you have received it. Follow your company's policy regarding this issue. A copy of the report will then be given to you or the person in charge at the time of the inspection. Copies of all reports should be kept on file in the operation. Copies of reports are also kept on file at the regulatory authority. They are considered public documents that may be available to anyone upon request.

Acting on all deficiencies noted in the report

You must make corrections within the timeline given by the inspector. Review your inspection report. Determine why the deficiencies happened by reviewing standard operating procedures. The master cleaning schedule, employee training, and foodhandling practices also should be reviewed, as shown in the photo at left. Revise current procedures or set up new ones to correct problems permanently. Inform employees of any deficiencies, and retrain them if necessary.

Closure of the Operation

Your operation can be closed for any of the following health hazards.

- Significant lack of refrigeration
- Backup of sewage into the facility or its water supply
- Emergency, such as a fire or flood
- Significant infestation of insects or rodents
- Long interruption of electrical or water service
- Clear evidence of a foodborne-illness outbreak

An inspector can suspend your permit to operate or ask you to close voluntarily. If an operation is suspended, it must stop operating immediately. However, an operation's owner can ask for a hearing if he or she believes the suspension was not justified. There may be a time limit to ask for a hearing. Check with your local regulatory authority.

Depending on the area, a suspension order may be posted at a public entrance to the operation, as shown in the photo at left. However, a posting is not usually required if the operation closes voluntarily.

To reinstate a permit to operate, the operation must get rid of the hazards that caused the suspension. It must then pass a reinspection.

Self-Inspections

Well-managed operations have frequent self-inspections to keep food safe. These are done in addition to regulatory inspections. In fact, regulatory inspections should be a supplement to self-inspections. They can be conducted in house or by a third-party organization.

A good self-inspection program provides many benefits.

- Safer food
- Improved food quality
- Cleaner environment for employees and customers
- Higher inspection scores

Strive to exceed the standards of the local regulatory authority. This will help you perform well on regulatory inspections. Also, your customers will see your commitment to safe dining experiences.

Consider the following guidelines when conducting a self-inspection.

- Use the same type of checklist that the regulatory authority uses.
- Start the inspection outside the operation and then proceed inside.
- Identify risks to food safety in your operation. One way to do this is to use the Food Safety Self-Assessment found in the appendix of this book.
- After the inspection, meet with staff to review any problems, as shown in the photo at left.

Apply Your Knowledge

The Inspection Process

Place a ✔ next to things you should do during a regulatory inspection.

① _____ Discuss violations and time frames for correction with the inspector.

② _____ Keep the relationship professional.

③ _____ Ask for identification.

④ _____ Refuse entry to an unexpected inspector.

⑤ _____ Correct the deficiencies you can immediately.

⑥ _____ Tell your employees not to answer any questions.

⑦ _____ Refuse to sign the inspection report.

⑧ _____ Take notes.

⑨ _____ Cooperate.

⑩ _____ Ask why a requested record is needed.

For answers, please turn to page 13.10.

Chapter Summary

The *FDA Food Code* outlines the federal government's recommendations for food safety regulations for the foodservice industry. Each state decides whether to adopt the *FDA Food Code* or some modified form of it. Enforcement is usually carried out at the state and local levels.

The inspection process lets an operation know how well it is following food safety practices. During an inspection, cooperate with the health inspector. Keep the relationship professional. Go with the inspector on the inspection, and take notes about any problems. If a problem can be corrected right away, do so. The inspector may ask for records, so be prepared to provide them. Discuss violations and time frames, and then follow up.

Operations with high standards for food safety consider regulatory inspections only a supplement to their own self-inspection programs. Self-inspections help an operation provide safer food and a cleaner environment for both staff and customers, in addition to higher regulatory inspection scores.

Chapter Review Activity

Now take what you have learned in this chapter and apply it to the following questions.

① Frank is getting ready for a regulatory inspection of his independent deli. Who will inspect his operation? Why?

② A manager was asked by her staff why regulatory inspections are important. How should she answer?

③ List at least three reasons why an inspector may suspend an operation's permit.

④ If an operation frequently receives regulatory inspections, does it need a self-inspection program? Why?

⑤ List three types of documents that an inspector may ask for.

⑥ If an inspection report notes a deficiency, when should a corrective action be taken?

For answers, please turn to page 13.10.

Study Questions

Circle the best answer to each question below.

① **A backup of raw sewage and significant lack of refrigeration can result in**

   A a delay of an inspection until the situation is corrected.

   B closure of the operation by the regulatory authority.

   C improved inspection scores.

   D being issued a permit to operate.

② **A person shows up at a restaurant claiming to be a health inspector. What should the manager ask for?**

   A Inspection warrant

   B Inspector's identification

   C Hearing to determine if the inspection is necessary

   D One-day postponement to prepare for the inspection

③ **Which agency enforces food safety in a restaurant?**

   A Centers for Disease Control and Prevention

   B Food and Drug Administration

   C State or local regulatory authority

   D U.S. Department of Agriculture

④ **Who is responsible for keeping food safe in an operation?**

   A Food and Drug Administration

   B Health inspectors

   C Manager/operator

   D State health department

For answers, please turn to page 13.10.

Answers

13.3   Who's in Charge Here?

① B   ④ E

② C   ⑤ D

③ D

13.7   The Inspection Process

1, 2, 3, 5, 8, 9, and 10 should be marked.

13.8   Chapter Review Activity

① The local government will inspect Frank's operation. While the FDA writes the *FDA Food Code*, local regulatory authorities are responsible for enforcing food safety regulations.

② The manager should tell her staff that regulatory inspections help make sure the operation is meeting minimum food safety standards. They are a chance for inspectors to provide food safety information to managers. Inspections also provide a written report of any food safety deficiencies.

③ An operation's permit can be suspended for any of these reasons.

- Significant lack of refrigeration

- Backup of sewage into the operation or its water supply

- Emergency, such as a fire or a flood

- Significant infestation of insects or rodents

- Long interruption of electrical or water service

- Clear evidence of a foodborne-illness outbreak related to the operation

④ Yes, an operation needs a self-inspection program. A good self-inspection program will help keep food safer than relying on regulatory inspections alone. Self-inspections also help an operation perform better on regulatory inspections.

⑤ An inspector can ask for any of these documents.

- Purchasing records

- HACCP records

- Proof of food safety knowledge

- Pest control treatments

- List of chemicals used in the operation

⑥ An operation must make a correction action within the timeline given by the inspector.

13.9   Study Questions

① B   ③ C

② B   ④ C

Notes

14

Employee Food Safety Training

In the News

Shedding Light on Dirty Hands

A local restaurant group has adopted a new method for teaching its staff about good personal hygiene. Each year, the organization holds a competition among staff members to see who has the best handwashing technique. The competition shows how dirty a person's hands can be, even after what seems like a thorough washing.

First, the staff puts a solution on their hands that makes dirt glow under a black light. Then they wash their hands. When the competitors do not wash their hands the right way, the chemical still appears around their cuticles, under their fingernails, and between their fingers. At the first competition, 90 percent of the staff got a failing score. But with guidance and commitment from management, scores have improved every year. Currently, even the local health department participates in the event.

You Can Build on This

The operation in the story above found a clever way to train staff on the right way to wash hands. The training showed one of the most important ways to prevent contamination of food. Using a competition to teach staff also made it fun and interesting. Like the operation above, you can be successful in teaching your staff food safety knowledge and skills. In this chapter, you will learn about the following training topics.

- Making sure that staff is trained initially and on an ongoing basis
- Identifying specific training needs
- Identifying tools for food safety training
- Keeping food safety training records

Concepts from Earlier Chapters

Before reading this chapter, remember these concepts and facts.

None

Food Safety Training for Staff

You have no guarantee of how long staff members will be working at your operation. Whether your staff has been on the job for one day or five years, they must understand that food safety is always important. To make sure your operation is serving safe food, you must train your staff when they are first hired and on an ongoing basis.

Training Staff

As a manager, it is your responsibility to make sure that your staff knows how to handle food safely. You must also tell them about updates to foodservice regulations, changes in the science of food safety, and new best practices in the industry.

Your first task is to identify the training needs in your operation. A training need is a gap between what staff needs to know to perform their jobs and what they actually know. For new hires, the need might be apparent. For experienced staff, the need is not always as clear.

Identifying your staff's food safety training needs will require effort on your part. However, there are several ways to do this, including the following ideas.

- Observing performance on the job
- Testing food safety knowledge
- Identifying areas of weakness

Your entire staff needs general food safety knowledge. Other knowledge will be specific to the tasks performed on the job. For example, everyone needs to know the right way to wash their hands. However, as shown in the photo at left, only receiving staff needs to know how to inspect produce during receiving.

Critical Food Safety Knowledge

You cannot assume that new hires will understand your operation's food safety procedures without training. From their first day on the job, they should learn about the importance of food safety and receive training in the critical areas listed below.

Personal hygiene

- Behaviors that can contaminate food
- Hand issues, which include how and when to wash hands, how to use hand antiseptics, how to use single-use gloves, when to change gloves, and correct hand care
- Personal cleanliness
- Correct work attire
- Health issues that must be reported
- Policies for eating, drinking, smoking, and chewing gum or tobacco
- Storage of dirty and contaminated clothing

Safe food preparation

- Preventing time-temperature abuse
- Identifying types of contaminants
- Learning how contamination occurs
- Preventing contamination and cross-contamination
- Handling food safely during the flow of food—receiving and storage, prepping and cooking, holding and cooling, and reheating and service
- Identifying common food allergens and methods for preventing allergic reactions

Cleaning and sanitizing

- Cleaning and sanitizing food-contact surfaces
- Identifying when cleaning and sanitizing are needed

Safe chemical handling

- Handling chemicals in the operation

Retraining

Your staff needs to be periodically retrained in food safety. You can retrain them by scheduling short training sessions, planning meetings to update them on new procedures, or holding motivational sessions that reinforce food safety practices.

Record Keeping

Keep records of all food safety training carried out at your operation. For legal reasons, when an employee completes this training, make sure to document it.

Apply Your Knowledge

Identifying Training Needs

For each of the following methods for identifying the training needs of your staff, list two ways you could use the method in your operation.

① Observe performance on the job.

_____

_____

_____

_____

② Test food safety knowledge.

_____

_____

_____

_____

③ Identify areas of weakness.

_____

_____

_____

_____

For answers, please turn to page 14.13.

Training Delivery Methods

There is more than one way to teach staff members what they need to know and do to keep food safe. As a manager, you must consider both the staff and the subject area you are teaching so that you can choose the best method.

When choosing training methods, think about what would work best in your operation. Some operations use a traditional method, such as on-the-job training (OJT). Others use a more activity-based approach. No single type of training works best, because everyone learns differently. Using many methods will provide the best results.

On-the-Job Training

Many operations use experienced staff members to teach learners while on the job, as shown in the photo at left. Learners repeatedly perform tasks while the trainers tell them how they are doing.

OJT teaches skills that require thinking and doing. It's good for training one staff member at a time, but it can also work for small groups. It's good for teaching skills that require watching someone do the task the right way.

Success depends on the ability and skill of the person doing the training. Therefore, you must choose the trainer carefully. Before using OJT, you should also recognize that it takes experienced staff away from their jobs. Additionally, it's not as effective for training large groups of people.

Classroom Training

Today's workforce expects training that will entertain and teach them. This can be challenging, but it's not impossible. Using an activity-based approach to training can be very effective. People learn by doing, instead of just being told what to do. Therefore, your training should include activities that require staff to do something.

Staff should also take part in learning activities. You must create a learning environment that encourages your staff to ask questions and allows them to make mistakes in that environment. You must also make your staff responsible for their own learning.

You can use many activity-based training methods to teach food safety to your staff.

- Information search
- Guided discussion
- Demonstration
- Role-play
- Jigsaw design
- Games
- Training videos and DVDs

Information Search

Some people are curious and like to explore things on their own. You can make use of their curiosity by having them find food safety information themselves rather than telling it to them. Here's how to do it.

1 Put staff in small groups.

2 Give them questions that they must answer in a set amount of time.

3 Give them the following types of tools to answer the questions.

- Operations manuals
- Job aids
- Posters, such as the one the manager in the photo at left is using
- Employee guides

4 Bring groups together and have them talk about what they learned.

Guided Discussion

Another way to teach food safety concepts is to ask your staff questions that draw on their knowledge and experience. Your goal is to make them think and discuss their thoughts. Each time learners answer a question, you should follow with another question.

Using this approach, a training session on calibrating thermometers might go something like this.

> *Instructor:* How can you find out if a cooked chicken breast has reached the right temperature?
>
> *Learner:* Use a thermometer.
>
> *Instructor:* How can you make sure a thermometer's reading is right?
>
> *Learner:* Calibrate it.
>
> *Instructor:* How do you calibrate a thermometer?
>
> *Learner:* By using the ice-point method or the boiling-point method.

Demonstration

Many times, you will teach specific food safety tasks by showing them to a person or group. Demonstrations are most effective when you follow the "Tell/Show/Practice" model. Here's how to do it.

❶ Tell

Tell the learner how to do the task. Explain what you are doing and why.

❷ Show

Show the learner how to do the task.

❸ Practice

Let the learner do the task. As extra practice, have the learner explain how to do the task before showing how to do it. Tell the learner how he or she is doing throughout the practice.

Role-Play

Many trainers use role-play to teach concepts. However, some learners don't like role-playing because it puts them on the spot. Role-play can work if you handle it the right way. Here's how to do it.

1. Prepare a script in advance that shows the right or wrong way to perform a skill.

2. Find two volunteers and give them time to rehearse the script. Do this early in the training session. As an alternative, the instructor can play one of the parts in the role-play.

3. Have the volunteers act out the script.

4. Ask the rest of the group to decide what the role-players did right and what they did wrong.

Jigsaw Design

There is an old saying that goes, "You've learned something when you can teach someone else how to do it." The jigsaw method follows this principle. Here's how to use it.

1. Put learners in small groups.

2. Assign a specific food safety topic to each group.

3. Tell each group to read about their topic, discuss it, and decide how to teach it to the other groups.

4. Take one person from each group and form new groups.

5. Have each member in the new group teach his or her topic to the other group members.

6. Bring the groups back together for review and questions.

Games

A game, as shown in the photo at left, can help make difficult or boring information seem more exciting. You can also use games to practice information that has already been taught. To be effective, games must meet the following criteria.

- Easy to play

- Fun

- Meets all time frames

- Easy to bring to the training site

- Easy to change for the audience and content

Training Videos and DVDs

In the training world, there is a general belief that learners remember the material in their training sessions in the following ways.

- 10 percent of what they read

- 20 percent of what they hear

- 30 percent of what they see

- 50 percent of what they see and hear

Using videos and DVDs, as shown in the photo at left, will help your staff see and hear their food safety training, making them more likely to remember it. Video is also very useful for teaching skills that involve motion, such as calibrating a thermometer.

If your staff is learning food safety on their own by video instruction, you should give them print materials as a supplement.

Technology-Based Training

Many operations use technology-based training to teach food safety, as shown in the photo at left. This includes online training and interactive CD-ROMs. Technology-based training lets you deliver training when and where your staff needs it. It's most appropriate in the following situations.

- Staff works in different locations and/or needs the same training at different times.

- It's costly to bring staff to the same place.

- Staff needs retraining to complete a topic.

- Staff has different levels of knowledge about a topic.

- Staff has different learning skills.

- Classroom training makes staff nervous.

- Staff needs to learn at their own pace.

- You want to collect specific information, such as time spent on different topics, test scores, number of tries until the training is finished, and/or problem areas.

Apply Your Knowledge

There's More Than One Way to Teach Your Staff!

Write the letter of the training delivery method on the line next to the statement describing the method.

Ⓐ Demonstration

Ⓑ Game

Ⓒ Guided discussion

Ⓓ Information search

Ⓔ Jigsaw design

Ⓕ On-the-job training

Ⓖ Role-play

Ⓗ Technology-based training

Ⓘ Videos and DVDs

① _____ Is best used when learners are located in different places and you want to collect information about their learning progress.

② _____ Allows learners to research a topic in a group and then teach the topic to a different group.

③ _____ Uses experienced staff members to teach a few learners food safety skills that require thinking and doing.

④ _____ Requires learners to explain how a food safety task should be performed before actually performing the task.

⑤ _____ Draws on learners' knowledge and experience by asking a series of questions about a topic.

⑥ _____ Allows learners to explore a topic on their own by using materials supplied by the instructor. Then they answer questions in a set amount of time and report their answers to the group.

For answers, please turn to page 14.13.

Chapter Summary

As a manager, you must ensure that your staff has the knowledge and skills needed to handle food safely in your operation. First, assess the training needs in your operation. A training need is a gap between what staff needs to know to do their jobs and what they actually know. There are a few ways to identify food safety training needs. You can test your staff's knowledge. You can also observe their performance. Another way is to survey them to identify areas of weakness.

All staff needs general food safety knowledge. Other knowledge will be specific to their jobs. Regardless of their positions, staff needs to be retrained on a regular basis. Keep records of all training conducted in your operation. For legal reasons, you need to document that staff has completed food safety training.

You can use many methods to train. No single method is best for all staff. This is because each person learns differently. Using many methods will result in more effective learning.

Chapter Review Activity

Now take what you have learned in this chapter and apply it to the following activity. Review the situations below and then answer the questions that follow each one.

① Next week, dishwashers and busers will begin using a different sanitizer from a new supplier. They are currently using quats for all their sanitizing duties.

What training method or methods would work best? Why?

_____

_____

② Your operation was recently inspected. The health inspector mentioned on the report that the food slicer is not being cleaned and sanitized the right way.

What training method or methods would work best? Why?

_____

_____

③ You have a new buser in your operation who has never bused tables before.

What training method or methods would work best? Why?

_____

_____

④ Last week, Chris hired Shauna, an experienced dishwasher. On Shauna's first day, Chris spent half an hour showing Shauna around the operation, including where to find the supplies for the machine. Because Shauna is an experienced dishwasher, Chris left her at the machine to start working. Although Shauna was experienced, she had never used this make of dishwasher. She was not quite sure how to load this machine for best results, but she did the best she could. When unloading the dishwasher, Shauna found that some of the dishes were still dirty. She looked for Chris to find out what she did wrong.

A What did Chris do wrong?

_____

_____

_____

_____

B How could Chris have helped Shauna be more successful at her new job?

_____

_____

_____

_____

For answers, please turn to page 14.13.

Study Questions

Circle the best answer to each question below.

① **When should staff receive food safety training?**

A When an employee is hired, and then periodically after that

B Only when hiring a new employee without foodservice experience

C When a new *FDA Food Code* comes out

D Only when they request it

② **New employees must be trained in the critical areas of personal hygiene, safe food preparation, cleaning and sanitizing, and**

A crisis management.

B equipment handling.

C HACCP plan creation.

D safe chemical handling.

③ **The manager's responsibility for staff food safety training is to**

A test staff's food safety knowledge.

B provide all staff with videos and DVDs for training.

C make sure that staff has the knowledge and skills to keep food safe.

D send all staff to a ServSafe training class.

④ **All new staff should receive training on**

A HACCP.

B crisis management.

C personal hygiene.

D active managerial control.

⑤ **What is the first task in training a large group of servers to prevent contamination of food?**

A Assess the training needs of the servers on this topic.

B Make a list of possible information to cover.

C Provide servers with operation manuals on the topic.

D Put staff into small groups, based on service experience.

⑥ **In which training method does a trainer ask a series of questions to draw on the knowledge and experience of the learners?**

A Information search

B Guided discussion

C Jigsaw design

D Games

For answers, please turn to page 14.13.

Answers

14.4 Identifying Training Needs

① Students' answers will vary, but they should focus on actions that allow them to watch employees completing food safety-related tasks. Examples include watching how someone takes the temperature of food or watching how someone washes his or her hands.

② Students' answers will vary, but they should focus on actions that have employees demonstrating or explaining their food safety knowledge. Examples include giving employees quiz sheets or asking employees to explain or show how to do a food safety task.

③ Students' answers will vary, but they should focus on documentation that could be used to check employees' areas of food safety weakness. Examples include reviewing temperature logs and self-inspection reports or implementing a mystery shopper program that highlights employees' food safety behavior.

14.10 There's More Than One Way to Teach Your Staff!

① H	③ F	⑤ C
② E	④ A	⑥ D

14.11 Chapter Review Activity

① On-the-job training and demonstration would be the best training methods, because staff can see and hear the right way to use the new chemical and can practice using it safely.

② Demonstration would be the best training method, because the manager can observe the staff member practicing the procedure.

③ On-the-job training would be the best training method for the new buser, because the new hire can see an experienced buser perform the job and then can practice with a mentor.

④ A Here is what Chris did wrong.

- He did not identify any training needs that Shauna had.

- He did not provide any general food safety training. He did not observe Shauna's work to determine what training she needed.

- He also did not provide specific training on the equipment Shauna needed to use.

B Here is what Chris should have done.

- He should have determined if Shauna had experience with the dishwasher she needed to use.

- He should have had an experienced dishwasher demonstrate how to use the machine and answer Shauna's questions.

- Chris or Shauna's supervisor should have observed her performance on the job and retrained her, if needed.

14.12 Study Questions

① A	③ C	⑤ A
② D	④ C	⑥ B

A

Appendix

Implementing Food Safety Practices Learned in the ServSafe Program

The ServSafe program will give you the information you need to keep food safe in your operation. It is your responsibility to put that information into practice. To do this, you must take what you have learned and use it to examine the following parts of your operation.

- Current food safety policies and procedures
- Employee training
- Your facility

The steps listed below will help you make the comparison that will take you from where you are today to where you need to be to *consistently* keep food safe in your operation.

❶ **Evaluate your current food safety practices using the Food Safety Evaluation Checklist in the appendix.** This checklist, which begins on page A.2, identifies the most critical food safety practices an operation must follow. It is a series of Yes/No questions that will help you see areas for improvement. When you have checked a "No" in this checklist, you have found a gap in your food safety practices. These gaps are the starting point for improving your current food safety program.

❷ **Review the "How This Relates to Me" areas throughout *ServSafe Essentials*.** These are the write-in areas in the book that help you remember the food safety practices required by your local regulatory authority. If a requirement is different from your company policy or is not addressed by it, you have found a gap in your food safety program and a chance to make an improvement.

❸ **Determine the cause of the gaps you found in steps 1 and 2.** For example, if you find that your walk-in cooler cannot hold food at 41°F (5°C) or lower, you have found a gap. There are many things that could have caused this gap, including faulty equipment, a walk-in door that is opened too often, etc. You must look at each of these potential causes to determine the true reason for the gap.

❹ **Create a solution that closes the gaps.** Your solution might include any of the following tasks.

- Developing or revising standard operating procedures (SOPs)
- Making improvements to current equipment or buying new equipment
- Training or retraining staff

❺ **Evaluate your solution regularly to make sure it has closed the gaps you found in steps 1 and 2.**

Food Safety Evaluation Checklist

The following self-assessment (pages A.2 through A.6) can help you find food safety gaps in your operation. It can also help you put food safety systems, such as active managerial control, into place. This self-assessment can help you address the five risk factors identified by the Centers for Disease Control and Prevention (CDC), as well as other food safety risks in your operation. It will also help you develop SOPs, policies, and useful food safety programs. Once you are finished, you can prioritize your gaps and work on creating a solution.

Directions

Check Yes after each question if your operation already performs the practice. Check No if it does not. Each "No" identifies a gap and offers a chance for revising your food safety program.

Failing to Cook Food Adequately; Holding Food at the Wrong Temperature

Topic/Principle	Evaluation	Page Reference in *Essentials*
① Are time and temperature controls part of every employee's job?	☐ Yes ☐ No	**5.4**
② Are time and temperature controls incorporated in your SOPs?	☐ Yes ☐ No	**5.4**
③ Are calibrated thermometers available to all foodhandlers?	☐ Yes ☐ No	**5.8**
④ Do you calibrate thermometers regularly?	☐ Yes ☐ No	**5.9 to 5.10**
⑤ Do all employees know how to use thermometers?	☐ Yes ☐ No	**5.6 to 5.11**
⑥ Do you minimize the amount of time food spends in the temperature danger zone (41°F to 135°F [5°C to 57°C])?	☐ Yes ☐ No	**5.4**
⑦ Do you document product temperatures in a temperature log or line check?	☐ Yes ☐ No	**5.4**
⑧ Do you reject food that has not been received at the right temperature?	☐ Yes ☐ No	**6.4 to 6.6**
⑨ Do you store TCS food at its required storage temperature?	☐ Yes ☐ No	**6.9 and 6.10**
⑩ Do you thaw food correctly?	☐ Yes ☐ No	**7.2**
⑪ Do you cook TCS food to the right minimum internal temperature?	☐ Yes ☐ No	**7.9**
⑫ Do you cool cooked TCS food according to the required time and temperature requirements?	☐ Yes ☐ No	**7.14**
⑬ Do you reheat TCS food that will be hot-held to 165°F (74°C) for fifteen seconds within two hours?	☐ Yes ☐ No	**7.16**
⑭ Do you hold TCS food at the right temperature (41°F [5°C] or lower or 135°F [57°C] or higher)?	☐ Yes ☐ No	**8.2**

Protecting Food and Equipment from Contamination

Topic/Principle	Evaluation	Page Reference in *Essentials*
① Do your handwashing stations have the necessary tools and supplies?	☐ Yes ☐ No	**10.7**
② Is the equipment you purchase designed with food safety in mind?	☐ Yes ☐ No	**10.5**
③ Do your employees store cleaning towels in a sanitizer solution between uses?	☐ Yes ☐ No	**11.13**
④ Do your employees know how often to clean and sanitize food-contact surfaces?	☐ Yes ☐ No	**11.3**
⑤ Do your employees know how to use the sanitizer in your operation?	☐ Yes ☐ No	**11.5**
⑥ Do your dishwashing employees know how to use the dishwashing machine?	☐ Yes ☐ No	**11.7 and 11.8**
⑦ Do your dishwashing employees know how to clean and sanitize items in a three-compartment sink?	☐ Yes ☐ No	**11.9**
⑧ Do your employees know how to clean nonfood-contact surfaces?	☐ Yes ☐ No	**11.12**
⑨ Do your employees know how to store clean and sanitized utensils, tableware, and equipment?	☐ Yes ☐ No	**11.10**
⑩ Do you have a master cleaning schedule in place?	☐ Yes ☐ No	**11.15**
⑪ Do you store food in a way that prevents contamination?		
A Do you store food in designated storage areas only?	☐ Yes ☐ No	**6.9**
B Do you store ready-to-eat food above raw meat, poultry, and fish?	☐ Yes ☐ No	**6.11**
C Do you store dry food away from walls and at least six inches (15 centimeters) off the floor?	☐ Yes ☐ No	**6.11**
⑫ Do you prep food in a way that prevents contamination?		
A Is the workflow of your operation designed for food safety?	☐ Yes ☐ No	**10.2**
B Do you assign specific equipment to each type of food product used in your operation?	☐ Yes ☐ No	**5.3**
C Do you clean and sanitize all work surfaces, equipment, and utensils after each task?	☐ Yes ☐ No	**5.3**
D When using the same prep table to prep food, do you prep raw and ready-to-eat food at different times?	☐ Yes ☐ No	**5.3**
E Do you use ingredients that need minimal preparation?	☐ Yes ☐ No	**5.3**
⑬ Do you hold food in a way that prevents contamination?		
A Do you shield or cover food to protect it from contamination?	☐ Yes ☐ No	**6.9 and 6.11, 8.2**
B Do you throw out food being held for service after a predetermined amount of time?	☐ Yes ☐ No	**8.2**

Continued on the next page ▶

► *Continued from previous page*

Topic/Principle	Evaluation	Page Reference in *Essentials*
⑭ Do you serve food in a way that prevents contamination?		
A Do you minimize bare-hand contact with ready-to-eat food?	☐ Yes ☐ No	**8.5**
B Do servers avoid handling the food-contact surfaces of glassware, dishes, and utensils?	☐ Yes ☐ No	**8.6**
C Do you maintain self-service areas in a way that prevents contamination?	☐ Yes ☐ No	**8.7 and 8.8**
⑮ Do you handle chemicals in a way that prevents contamination?		
A Do you store chemicals away from food, utensils, and equipment?	☐ Yes ☐ No	**11.14 and 12.8**
B Are containers used to dispense chemicals labeled?	☐ Yes ☐ No	**11.14 and 12.8**
C If pesticides are used in the operation, are all food and food-contact surfaces removed prior to use?	☐ Yes ☐ No	**12.7**
⑯ Do you use only food-grade utensils in your operation?	☐ Yes ☐ No	**3.2**
⑰ Is your lighting installed in a way that does not contaminate food?	☐ Yes ☐ No	**10.10**

Using Approved Suppliers

Topic/Principle	Evaluation	Page Reference in *Essentials*
① Do you purchase food from suppliers that get their products from approved sources?	☐ Yes ☐ No	**6.2**
② Do you make sure that your suppliers are reputable?	☐ Yes ☐ No	**6.2**
③ Do your suppliers deliver during off-peak hours?	☐ Yes ☐ No	**6.2**

Good Personal Hygiene

Topic/Principle	Evaluation	Page Reference in *Essentials*
① Are all employees aware of how they can contaminate food?	☐ Yes ☐ No	**4.2 and 4.3**
② Do all employees follow the right procedure for handwashing?	☐ Yes ☐ No	**4.5**
③ Are all employees aware of when handwashing is required?	☐ Yes ☐ No	**4.6**
④ Do all employees follow hand maintenance procedures, such as keeping nails short and clean, and covering cuts and sores?	☐ Yes ☐ No	**4.7**
⑤ Do you provide the right type of gloves in your operation for handling food?	☐ Yes ☐ No	**4.8**
⑥ Do employees change gloves when necessary?	☐ Yes ☐ No	**4.9**
⑦ Do you have requirements for work attire for foodhandlers?	☐ Yes ☐ No	**4.10**
⑧ Do you require employees to maintain personal cleanliness?	☐ Yes ☐ No	**4.9**

Topic/Principle	Evaluation	Page Reference in *Essentials*
⑨ Do you prohibit employees from smoking, eating, or drinking in food-prep and dishwashing areas?	☐ Yes ☐ No	**4.11**
⑩ Do you have policies to address employee illnesses?	☐ Yes ☐ No	**4.12**
⑪ Do you model proper foodhandling behaviors at all times?	☐ Yes ☐ No	**4.4**

Facilities and Equipment

Topic/Principle	Evaluation	Page Reference in *Essentials*
① Is stationary food equipment installed the right way?	☐ Yes ☐ No	**10.6**
② Does your food equipment receive regular maintenance?	☐ Yes ☐ No	**10.6**
③ Is your plumbing installed and maintained by a licensed plumber?	☐ Yes ☐ No	**10.8**
④ Is lighting set at intensities that ensure food safety?	☐ Yes ☐ No	**10.10**
⑤ Is garbage removed from the premises correctly?	☐ Yes ☐ No	**10.11**

Pest Control

Topic/Principle	Evaluation	Page Reference in *Essentials*
① Do you have a contract with a licensed pest control operator?	☐ Yes ☐ No	**12.6**
② Do you inspect deliveries for signs of pests?	☐ Yes ☐ No	**12.2**
③ Do you take measures for preventing pests from entering the operation?	☐ Yes ☐ No	**12.2 and 12.3**
④ Do you take measures for denying pests food and shelter in the operation?	☐ Yes ☐ No	**12.4**
⑤ Can your employees identify signs of pests?	☐ Yes ☐ No	**12.6 and 12.7**

Food Safety Systems

Topic/Principle	Evaluation	Page Reference in *Essentials*
① Do you have prerequisite food safety programs in place?	☐ Yes ☐ No	**9.2**
② Does your food safety management system focus on controlling the CDC's five most common risk factors responsible for foodborne illness?	☐ Yes ☐ No	**9.3**
③ Does your food safety management system focus on identifying, monitoring, and controlling biological, chemical, and physical hazards?	☐ Yes ☐ No	**9.7**
④ Do you know when a HACCP plan is required?	☐ Yes ☐ No	**9.12**

Employee Training

Topic/Principle	Evaluation	Page Reference in *Essentials*
① Do you have food safety training programs for both new and current employees?	☐ Yes ☐ No	**14.3 and 14.4**
② Do you have assessment tools that identify food safety training needs for employees?	☐ Yes ☐ No	**14.2**
③ Do you have a variety of food safety training resources (including books, videos, posters, and technology-based materials) to meet your employees' learning needs?	☐ Yes ☐ No	**14.5 to 14.9**
④ Do you keep records documenting that employees have completed training?	☐ Yes ☐ No	**14.4**

Auditing (Self-Inspection)

Topic/Principle	Evaluation	Page Reference in *Essentials*
① Do you perform regular self-inspections?	☐ Yes ☐ No	**13.7**
② Do you regularly compare your local or state food safety regulations to procedures at your operation?	☐ Yes ☐ No	**13.7**
③ Are all infractions from regulatory inspections or self-inspections taken care of in a timely manner?	☐ Yes ☐ No	**13.6**
④ Do you have a plan for working with health inspectors during inspections?	☐ Yes ☐ No	**13.5 and 13.6**

Notes

_____
_____
_____
_____
_____
_____
_____
_____
_____
_____
_____
_____
_____
_____
_____
_____
_____
_____
_____
_____
_____
_____
_____
_____
_____
_____
_____
_____
_____
_____
_____
_____
_____
_____

I

Index

Notes